Becoming a Scientist in Mexico

Becoming

Scientist in Mexico

The Challenge of Creating
a Scientific Community
in an Underdeveloped Country

Jacqueline Fortes and Larissa Adler Lomnitz
Translated by Alan P. Hynds

The Pennsylvania State University Press
University Park, Pennsylvania

Library of Congress Cataloging-in-Publication Data

Fortes, Jacqueline.
 [Formación del científico en México. English]
  Becoming a scientist in Mexico : the challenge of creating a scientific community in an underdeveloped country / Jacqueline Fortes and Larissa Adler Lomnitz ; translated by Alan P. Hynds.
    p.   cm.
  Translation of : Formación del científico en México.
  Includes bibliographical references and index.
  ISBN 0-271-01018-5 (acid-free paper)
  1. Science—Social aspects—Latin America. 2. Scientists—Training of—Mexico. 3. Scientists—Social aspects—Mexico—Case studies. 4. Universidad Nacional Autónoma de México. I. Lomnitz, Larissa Adler de. II. Title.
Q175.M6F6713   1994
306.4'5'0972—dc20   93-15760
                                                                  CIP

Copyright © 1994 The Pennsylvania State University
All rights reserved
Printed in the United States of America

Published by The Pennsylvania State University Press,
Barbara Building, Suite C, University Park, PA 16802-1003

It is the policy of The Pennsylvania State University Press to use acid-free paper for the first printing of all clothbound books. Publications on uncoated stock satisfy the minimum requirements of American National Standard for Information Sciences—Permanence of Paper for Printed Library Materials, ANSI Z39.48–1984.

To Enrique, my companion of always
To Sergio and Tatiana, my beloved children
—Jacqueline Fortes

To my dear children, Jorge, Claudio, Alberto, and Tania,
and grandchildren, Enrique, Elisa, Michael, and Jason
—Larissa Adler Lomnitz

Contents

Acknowledgments	ix
Introduction	1
1 THE DEVELOPMENT OF SCIENCE AND THE UNIVERSITY OF MEXICO, 1551–1980 Science in Mexico; The Latin American University; The National Autonomous University of Mexico	11
2 ORIGINS AND DEVELOPMENT OF THE FIELD OF BASIC BIOMEDICAL RESEARCH The Origins of the New Undergraduate Program; Internal Structure of the Undergraduate Program; The Development of the Undergraduate Program, 1974–1980; Work Routine; Academic Traits of the Teachers; The First Year	31
3 EXPERIMENTAL RESEARCH WORK: THE TEACHING METHODOLOGY Learning Manual Skills; Problem Solving; Learning to Read and to Question; Learning to Identify and Solve Problems; Working in Research; Discussion	55
4 IDEOLOGY AND SOCIALIZATION: THE IDEAL SCIENTIST Control Structure; Liberating Processes; Scientific Ideology in the Literature of Sociology	75
5 THE SOCIALIZATION PROCESS The Initial Contact; The Relationship with the Teacher/Advisers; Interaction with the Group; Role-Playing; Socialization in the Third and Fourth Years	97

6 THE ACQUISITION OF A SCIENTIFIC IDENTITY
   Developing a Scientific Identity; Stages in the Process of
   Identification; Discussion; Conclusions   143

Appendix 1: Curriculum of the Undergraduate Program in Basic
   Biomedical Research, First and Third Cohorts   165

Appendix 2: The Formal Knowledge That Was Taught   169

Appendix 3: The Students' Psychological Traits   183

Bibliography   211

Index   219

Acknowledgments

We wish to thank Drs. Jaime Mora, Jaime Martuscelli, and Kaethe Willms, who, in their capacity as assistant directors of the Instituto de Investigaciones Biomédicas of the Universidad Nacional Autónoma de México (UNAM), supported this project. We also express our gratitude to the Instituto's faculty and to the undergraduate students in basic biomedical medicine. Periodically agreeing to talk with us for long hours, they rendered invaluable assistance.

An earlier version of this work was published in Spanish, under the title *La formación del científico en México: Adquiriendo una nueva identidad* (México: Siglo XXI, 1991).

Introduction

The purpose of this book is to contribute to the social and cultural understanding of science in Latin America and, more specifically, to study the problem of the training and development of scientists in Third World countries. We draw on sociology, anthropology, and psychology in order to analyze a case study on the socialization of scientists in Mexico.

It might be felt, initially, that these two projects are independent from one another. Indeed, the issue of science and its development in countries that arrive late to the industrialization process can be dealt with separately from the problem of transforming secondary-education graduates into scientific researchers. The latter problem is not unique to the periphery: it can be observed everywhere, and differences need not arise by dint of a national society's level of development. Nevertheless, it is a fact that socialization of scientists in peripheral countries occurs under adverse conditions as compared to the conditions that prevail in countries where science responds to its own economic, social, and cultural traits.

The undergraduate program in basic biomedical research of the National Autonomous University of Mexico (UNAM) represents, in our opinion, a crucial experiment for solving the problem of training professional scientists in Mexico. There, as in many Latin American countries, university teaching has traditionally focused on forming professionals who use knowledge, rather than researchers who produce it. Teaching and research have been separate areas: there are departments (called *facultades*), which are dedicated to teaching and training; and there are institutes, devoted exclusively to research. It is necessary to keep this institutional framework in mind if we are to appreciate the groundbreaking and innovative nature of the teaching experiment analyzed in this book.

Mexico is a developing society that is beginning to perceive the importance of technology. Still, until recently, there had not been a clear notion regarding the role of science or the scientist, nor of the exact nature of the country's incipient scientific community. Given this historical and cultural framework, it is natural that scientific careers still lack the sort of recognition traditionally enjoyed by such professional careers as medicine, law, and engineering.

The Latin American educational tradition has given little importance to the task of encouraging the qualities necessary to a scientific mind. The central thesis of this book is that the transmission of scientific ideology (or ethos) is the key link in training researchers. Knowledge and techniques are necessary but insufficient conditions in training scientists; ideological aspects (i.e., beliefs and values) hold a predominant position. Therefore, our research focused on understanding the ideal model that underlay the socialization process we studied and how that model was transmitted to and assimilated by students.

In the peripheral countries, the training of scientists faces institutional and cultural conditions different from those that exist in the central countries where Western science arose (Gaillard 1991). These conditions, which we describe in Chapters 1 and 2, can be understood as unfavorable for the process of training researchers; that is to say, for encouraging their development. This fact defines both the context in which the undergraduate program being studied has developed and the characteristics which that program has acquired in relation to its organization, isolation, dedication, and internal demand for a level of excellence comparable to that of the central countries. What we are presenting here, then, is a study of ideology and socialization, within a framework of the historical development of science in Mexico up to the 1980s, and a study of the cultural and institutional context in which they occur.

Over the past decade, in light of the new international order, the importance of technology and of the production of knowledge has been underscored. A developing country such as Mexico now has the opportunity to participate with the developed countries on a basis of fairness. Thus the most recent proposal of the Economic Commission for Latin America and the Caribbean (ECLAC) stresses that human resources must be trained and that mechanisms favoring the generation of new knowledge must be encouraged and developed. In other words, the proposal highlights the systemic link between education, knowledge, and development (ECLAC-UNESCO 1992). In response to this situation, there has been a structural change in the Mexican economy in recent years, including the introduction of new economic and social measures, among them a program for scientific development through the National Council of Science and Technology (CONACYT). The Mexican state is making a strong effort to quickly develop a scientific base, through large support investments in research projects, scholarships, and graduate development programs that encourage the training of scientific researchers. In addition, in 1985 the National System of Researchers (SNI) was created, the task of which is to augment the salaries of researchers who practice within the country, in order to prevent "brain drain."

Tables 1–4 and Figures 1 and 2 show public expenditures for various development programs as well as the evolution of those expenditures during the past decade. We can see the increase in support given to research as well as in the support given to training human resources. However, financial support alone will not produce researchers. Pesos must be complemented by knowledge concerning the special problems entailed in training researchers in a country like Mexico as well as by a thorough understanding of the internal process of training scientists. Our case study sheds light on both of these critical areas.

This study covers the 1974–1980 period, during which the first three cohorts of the UNAM undergraduate program in basic biomedical research were formed. Our analysis focuses on diverse aspects of the transmission and internationalization of scientific ideology, within the context of a process of socialization and acquisition of a scientific identity.

Because of our interest in becoming thoroughly familiar with individual scientists, rather than in conducting a study concerned with large samples, which would necessarily sacrifice depth in favor of quantity of information, we have followed a qualitative methodology based on open-ended interviews and participatory observation. Over the course of the interviews our ideas gradually evolved and were partially transformed by the initial findings. The very topic of ideology arose from our interviews with the teachers, whose role in training a certain type of researcher was decisive and who were so intent on conveying scientific values to their students.

In all, 350 open-ended interviews were conducted during the six years

Table 1. Mexico's expenditures on scientific-technological development, by Program, 1980–1991 (millions of pesos)

			Promotion & Regulation of:				
	Administration	Planning	Scientific Research & Technological Development	Scientific & Technological Services	International Cooperation	Promotion of Training Human Resources	Total
1980	244	...	274	286	38	965	1,807
1981	501	10	594	423	62	1,467	3,057
1982	793	13	882	522	104	2,500	4,814
1983	855	28	2,032	352	160	3,668	7,095
1984	1,886	50	4,020	823	165	4,825	11,769
1985	2,712	80	8,424	1,156	230	6,674	19,276
1986	3,917	124	7,830	2,200	199	10,522	24,792
1987	9,925	226	16,740	4,305	380	21,463	53,039
1988	21,456	668	37,332	6,046	545	44,238	110,285
1989	24,359	843	46,660	7,328	720	49,263	129,173
1990	18,105	1,816	98,949	11,677	3,187	67,958	201,692
1991*	21,398	2,316	246,367	15,535	4,179	184,682	474,477

SOURCE: CONACYT-SEP 1991, 66.
*Preliminary data.

Table 2. Use of government resources, 1987–1991 (millions of pesos)

	1987	1988	1989	1990	1991*
Operating expenses	15,298	29,881	34,163	41,703	40,484
Support for National System of Science & Technology (SINCYT)	37,741	80,404	95,010	159,989	433,993
TOTAL	53,039	110,285	129,173	201,692	474,477

SOURCE: CONACYT-SEP 1991, 70.
*Preliminary data.

that the research project lasted. The thirty-one teachers and twenty students participating in the first four cohorts of the undergraduate program (including those who dropped out of the program during the early months) were interviewed repeatedly. For the last draft, though, we decided in favor of an analysis of the first and third cohorts, since only in the third was there a program change after the internal evaluation of the first two years. The interviews took place at the beginning and end of each semester, and the statements from those interviews were compared with our observations on everyday behavior and on such rituals as lectures, seminars, and congresses.

In each interview we asked the respondent to describe in detail the evolution of the seminars and to analyze the topics studied, the teaching methods used, the problems found with the program, and the difficulties faced by the teachers and the students. We asked each of the teachers to describe the objectives, methods, and development of their courses, including the difficulties they encountered and the changes they had to make. We also asked the teachers to evaluate each of the students, and we inquired into the grounds used in making those evaluations: which aspects and qualities stood out, and what parameters had been used?

Likewise, we asked all students to describe their activities, what they thought about what their teachers had tried to teach them, and how well their teachers had performed as well as what they felt they had assimilated. This, in turn, was compared with the teacher interviews and evaluations. In addition, we analyzed, at the beginning and end of each semester, all students' feelings and expectations as well as the image they had, at the time, of scientists.

With most of the teachers, we had long discussions concerning the role

Table 3. Support for the National System of Science and Technology (SINCYT), 1991 (millions of pesos)

Projects for scientific & technological development	81,862
National System of Researchers (SNI)	43,659
Scholarships within Mexico	35,741
Scholarships overseas	51,472
Support for graduate studies	12,000
Industry–academia liaison	5,080
International cooperation	4,179
Presidential funds	200,000
TOTAL	433,993

SOURCE: CONACYT-SEP 1991, 71.
NOTE: All data are preliminary.

and problems of scientists in Mexico. We also analyzed their teaching philosophy and methods as well as their expectations for their students, especially regarding scientific attitudes, cognitive abilities, and practices.

After each semester, we prepared a written description of the program's development, including topics covered, methodology, traits of the teachers and students, teacher–student interaction, specific problems, and evaluations by students and teachers. Later we developed a list of norms and values, and interpretations thereof, which we grouped according to similarity of function; some were explicitly expressed by teachers and/or students, and others were inferred from their comments. We analyzed the role of these values and described the mechanisms of their transmission and the difficulties that arose in their assimilation. We discussed our findings with some members of the Institute of Biomedical Research, and we interviewed in greater depth six of the teachers most active in the undergraduate program in order to obtain their comments on our work.

In this process, our model began to take shape. We abstracted an ideal model, following Weber's advocacy of "a unilateral stressing of one or more viewpoints [and a] synthesis of specific individual phenomena, many of which are diffuse, discrete, more or less present and occasionally absent, organized in a unified analytical construction" (1949, 90). In this way, we obtained an "ideal type" of the scientist who underlies this pedagogical experiment in undergraduate education. It is important to point out that "this ideal is not only an auxiliary logical instrument,

Table 4. Number of researchers in the National System of Researchers (SNI), by Research Area, 1984–1991

	Physico-Mathematical Sciences	Biological, Biomedical & Chemical Studies	Social Sciences & Humanities	Engineering & Technology	Total
1984	1,396	585	600	211	0
1985	2,276	859	970	447	0
1986	3,019	950	1,150	580	339
1987	3,458	757	1,100	699	902
1988	3,774	624	1,021	713	1,416
1989	4,666	718	1,237	855	1,856
1990	5,704	816	1,512	1,141	2,235
1991*	6,442	873	1,736	1,320	2,513

SOURCE: CONACYT-SEP 1991, 28.
*Preliminary data.

but a model in which respondents' value judgments were included" (Weber 1949, 97–98).

The difficulties faced in this type of study should be pointed out. On the one hand, there is greater emotional involvement when the researcher belongs to the same community that he or she is studying, and this implies a possible loss of objectivity.[1] There is, on the other hand, constant interaction between the ideal model and actual behavior, which often prevents a clear differentiation between the two, not to mention the mutual influence that takes place between researchers and study subjects, with the resultant modification of their conduct and interpretations. We do not claim to have resolved these difficulties to our complete satisfaction; but we do believe that, once the fieldwork was concluded and we had begun to analyze the data, we were able to establish a certain distance from the group being studied. For the entire period of the study, we had the cooperation of all respondents.

This book is made up of six chapters. Chapter 1 describes the

---

1. The difficulties for the researcher who would study the community to which he or she belongs are described by Pierre Bourdieu, who says: "When a research project studies the same territory as that in which one operates, the findings obtained can be immediately reversed in scientific work as instruments of reflection on the conditions and social limits on work, which represent one of the principal tools of epistemological surveillance" (1988, 15).

Fig. 1. Evolution of Mexico's expenditures on scientific-technological development, by program, 1985–1991

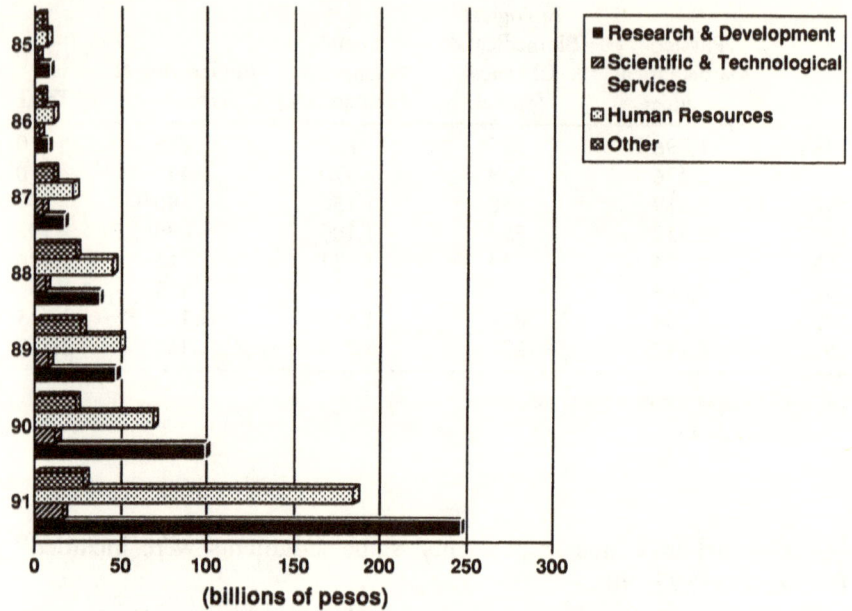

SOURCE: CONACYT-SEP 1991, 67. All data for 1991 are preliminary.
NOTE: "Other" = administration, planning, and international cooperation.

historical and institutional stage on which the study takes place, including a brief summary of the development of science in Mexico and the evolution of UNAM. Chapter 2 describes the undergraduate scientific program itself, from its antecedents until its present stage of development. Chapter 3 analyzes the teaching methodology. In Chapter 4 we submit the ideal model of the scientist that we developed from our observations in Chapter 5, which is a description and analysis of how the socialization process unfolded in the two cohorts under study. Finally, we conclude with an analytical discussion of the acquisition of a scientific identity.

Fig. 2. Support for the National System of Science and Technology (SINCYT), 1991

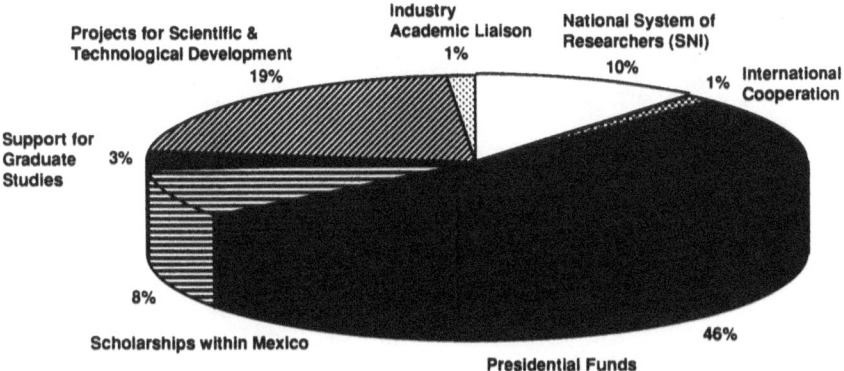

Source: CONACYT-SEP 1991, 71. All data are preliminary.

1

The Development of Science and the University of Mexico, 1551–1980

The modern world carries the mark of science; but science, in turn, responds to the requirements and obeys the limitations set by society. Science is a cultural product that emerged in the societies of Western Europe as the result of a complex historical evolution. It arose from a series of social developments such as the growth of cities, the decline of feudalism, the increase of commerce, the rise of the bourgeoisie, and the emergence of capitalist society.

During the European Renaissance, the idea that man, through his intelligence, can subdue and control the physical environment first received widespread currency. This idea implied the existence of a special type of knowledge, one that allowed the behavior of nature to be foreseen and predicted through the observation of its uniformities or regularities (Hull 1973, 224; Rose and Rose 1969).

The Renaissance spirit or mentality began to displace medieval European thought: speculative philosophy and Scholasticism. In his classic work on the Protestant ethic, Weber (1949) stresses the role of the religious Reformation as a determinant of the emergence of capitalism and Western science alike. The spirit of the new times was characterized by accounting and the use of technology for economic purposes. Calvinism especially, with its rational and empirical orientation, favored the development of experimental science. Thus, the new mentality tended to promote individualism, and it generated conflicts between reason and authority. During the Renaissance and in subsequent centuries, then, a new philosophy, intimately related to the development of science, gradually evolved and led, on the one hand, to a repudiation of the Greek ideas that had been incorporated into Scholasticism and that had to be submitted to rational analysis and, on the other hand, to a resurrection of the Greek spirit of free inquiry into nature and human reality (Hull 1973, 324).

The same period (1450–1690) that witnessed the development of capitalism as the dominant mode of production in Europe was also the period that developed differential calculus and experimental physics. Changing production techniques influenced science, which in turn produced accelerated changes in technology. According to Bernal (1979, 373), Europe's scientific-technological revolution represents a unique social phenomenon, for science won a permanent role among society's productive forces.

Historians of science usually focus their efforts on analyzing the intellectual and theoretical aspects of this great transformation. However, the originality of the new ideology of control over nature consisted precisely in establishing a pragmatic bridge between theory and practice. Observation and experimentation as praxis make modern science the powerful instrument of knowledge and action that it still is today.

Not in all societies did the conditions arise for the consolidation of modern science; for capitalism soon spread to the European colonies, and from then on there was a "center" and a "periphery" (Wallerstein 1974). In the central countries science prospered and continues to prosper: it was there that the material, social, and ideological conditions for its development arose, particularly in Northern Europe after the Reformation. In Spain, for example, the Counter-Reformation triumphed just as the spirit of the Renaissance was spreading throughout France and Italy. The colonization of the American continent allowed Spain to entrench itself against the rest of Europe and to become the stronghold of the Scholastic spirit and of an authoritarian and theocratic discourse that was on the way out elsewhere in Europe.

Science in Mexico

To New Spain the Spanish brought the Scholastic scientific culture that dominated Spain. In 1551, a few years after the *conquista*, the Real y Pontificia Universidad de la Nueva España was founded. Created and sustained by the Crown, administered by the clergy, it was the official university of the empire. The subjects taught, as well as their content, were determined by the various religious orders. Students were required to master a formidable set of theological subjects, regardless of the field they studied. It was an elitist university for forming educated *criollos* (creoles), but it also constituted an important factor in the development of Mexican *mestizo* (mixed-race) culture, since from the beginning it admitted the sons of Indian *caciques,* or chiefs. It was the hub of intellectual and religious development, and it became the most important cultural center in Latin America (Silva Herzog 1974; De María y Campos 1981; García Stahl 1975).

By the late eighteenth century, 1,162 doctors and 29,882 bachelors had graduated from the Pontifical University of Mexico, in addition to a large number of lawyers. Some eighty bishops had studied there. From the time it was founded, the university maintained a militant, anti-Reformation Scholastic spirit that prevailed for three centuries (De María y Campos 1981, 19–23; García Stahl 1975, 37–62; Cruz Manjarrez 1982, v–vii; De Gortari 1979, 204–5).

Initially, the university had two chairs, that of theology and that of the arts, which was divided into natural philosophy and moral philosophy. Natural philosophy included the study of the natural sciences, physics, mathematics, and geometry, all from the perspective of Aristotelian philosophy. In 1557, the Augustinian priest Alonso Gutiérrez wrote America's first scientific text, *Physica speculatio,* which contains the accepted doctrine on astronomy, meteorology, physics, botany, and psychology.

The Enlightenment

During the sixteenth and seventeenth centuries, while modern science was being developed in Europe, official doctrine in Mexico remained faithful to the texts of Aristotle and St. Thomas. The renowned Golden Age was nothing other than the flourishing of letters and fine arts that arose in Spain and Mexico around such traditional institutions as the university.

Science in colonial Mexico was a dependent science, based on European traditions and institutions. Perennially lagging behind science in France, England, or Italy, science in Spain fought a fruitless rearguard action. Although some exceptional individuals emerged in New Spain who did attempt to introduce modern ideas, science was neither promoted nor given continuity (Moreno de los Arcos 1975, 29–31).

When crisis finally broke out, "created by the absurd intromission of metaphysics in the field of the sciences" (Bravo Ugarte 1947, 229), there was in Mexico an anti-Spanish, creole reaction. The expulsion of the Jesuits, who had introduced the new ideas in Mexico, helped to antagonize the creoles and to promote a Mexican national feeling that was nourished by European science, by liberalism, and by the rejection of traditional philosophy.

Slighting the "Peripatetics," the new scholars quoted Galileo, Copernicus, Descartes, and Kepler. Their precursors in the eighteenth century were Don Carlos de Sigüenza y Góngora and Sor Juana Inés de la Cruz (De Gortari 1979, 225–30; Paz 1986). "The sciences, which until then had been scorned, began to be eagerly cultivated. People tried to become familiar with the geographic environment, the wealth and possibilities of what until then had been an unexplored continent. America attempted to discover itself, and in this task it was helped by botanists, astronomers, mineralogists, geographers, mathematicians. . . . Cartesianism served to advance the development of the sciences" (García Stahl 1975, 70). The savants of the eighteenth and early nineteenth centuries were intimately related to the university.

With the Bourbon reforms in Spain, important changes arose, including an attempt to catch up in science. This thrust also arrived in the colonies. Such authors as Descartes, Leibnitz, and Locke began to circulate surreptitiously. Travelers and political personages brought with them the influence of rationalist philosophical currents; the new ideas were discussed at social gatherings. The Jesuits, as noted, had already introduced educational renewal in their schools and had taught the philosophy and physics of the day. In Mexico, Francisco Javier Clavijero was the central figure of the "innovating movement," and he defended the need for the experimental method (Cruz Manjarrez 1982, 60–64).

The Enlightenment was a period of great literary and scientific effervescence. Astronomical and geographic observations were made; engineering projects and wildlife studies were conducted. Such figures as the Jesuit Andrés de Guevara, Joaquín Velázquez Cárdenas, José Alzate y Ramírez, J. I. Bartolache, Antonio León y Gama, and Andrés Manuel del Río appeared. These "savants" researched without laboratories or study centers; but around 1792 the Seminary of Mining was created, which would later become the College of Mining, in which Mexico's first modern physics laboratory was established. Courses were taught, textbooks were written, and the seminary became fertile ground for scientists (Cruz Manjarrez 1982, 66–68).

Science during the Mexican Enlightenment can be divided into four periods: the early period (1735–1767); the creole period (1768–1788); the official or Spanish period (1789–1803); and the period of synthesis (1804 until the beginning of the independence movement in 1810) (Moreno de los Arcos 1975, 25–41). After the 1767 expulsion of the Jesuits, the first scientists were self-taught creoles. Later on, this group was strengthened by the arrival of a group of Spanish scientists and by Humboldt's sojourn in Mexico. A new generation of peninsular Spaniards and creoles was devoted to research, teaching, and publishing and translating texts. Seminars were given in which the ideas of Bacon and Descartes were openly discussed, causing Scholastic philosophy, which continued to dominate the university, to lose strength. Mexican society prepared for independence (De Gortari 1979, 260–65). In sum, scientific development during the Enlightenment was related to the social, economic, and ideological changes that were occurring in Spain: the opening up to commerce, the Church's loss of power, and the increasing influence of the rest of Europe which would lead, belatedly, to Spain's industrial revolution.

Positivism and Revolution

The Wars of Independence brought an end to the Mexican Enlightenment and its scientific ambitions. For decades scientific activity in Mexico was almost nonexistent. In 1833 the Real y Pontificia Universidad was officially closed, implying a break with the monarchy and the Church. There followed a period of disorganization and of struggle between conservatives and liberals over who would lead the new nation.

Finally, the 1857 reform introduced the civil administration of justice, the establishment of free and mandatory primary education, and a new interest in higher education. Scientific institutes and societies were founded, such as the Humboldt Scientific Society (1862) and shortly after, the Chapultepec Astronomical Observatory, which Maximilian later ordered to be destroyed so that he could carry out the landscaping of his country castle. The observatory was rebuilt in 1878, when Mexico was invited to participate in the work of cataloging and photographically charting the heavens. Thus Mexico assumed its first international commitment, which placed it in the vanguard of astronomical science during this period (De Gortari 1979, 13–14; Cruz Manjarrez 1982, 112–14).

In the last decades of the nineteenth century, the industrialization process began in Mexico. This entailed an attempt to modernize and, at the same time, reinforced the underlying nationalistic political structure. A railroad was built and the government strove to unify the nation economically and politically. A strong boost was given to public education, under the influence of a group of thinkers who had embraced positivism and the scientific, progressivist faith.

Among these thinkers stands out Gabino Barreda, a student of Comte who in 1867 was charged with the commission aimed at reforming education. Algebra, geometry, calculus, physics, and chemistry were introduced into the secondary-school curriculum. The Escuela Nacional Preparatoria (National Preparatory School) was founded, guided by the precepts of Comte. Its teachers were positivists and began to dominate the intellectual climate of the era. Among them were such luminaries as Justo Sierra, Ezequiel Chávez, and Gabino Barreda himself. This group of young people organized and began to publish journals (De Gortari 1979, 303–4).

Not only did the positivists influence the new educational system, but their faith in science and progress led to a renaissance of scientific activity in Mexico. During the last years of Porfirio Díaz's rule the most influential sector in the management of economic policy was the so-called *grupo de los científicos,* who were the forerunners of what today we would call the technocracy but whose support for the dictatorship shrouded them in a sort of black legend. The target of all criticism before, during, and after the 1910 Revolution, the *científicos* came to monopolize the role of the villain in historiography on the Porfirian period (De María y Campos 1981, 3).

The word "science" came to the new postrevolutionary society worn out and surrounded by suspicion, for it had become identified with the exploitation and dictatorship that marked the rule of Porfirio Díaz (1876–1910). Although positivism had been important in the formation of the intellectuals of the Revolution, they gradually broke away from that system's philosophical rigidity. From within such organizations as Ateneo de la Juventud (Atheneum of Youth) they struggled for a greater cultivation of the humanities, for the "values of the spirit," and for a new encounter with Mexican reality and with Mexico's indigenous population. The intellectuals' banner was popular education: elementary rather than higher education. In revolutionary proclamations they demanded an end to the vestiges of *cientificismo;* no role was assigned to science for the task of social change.

In all of this we can see the relationship that exists in Mexico between the history of science and sociopolitical conditions. During the colonial period, science was an extension of the Spanish Crown's militantly anti-Reformation position. Later, in the eighteenth century, the Enlightenment emerged in response to Spain's need to come to terms with the European world. Then came independence, and the Enlightenment ended. Fifty years later, with the Reformation and incipient industrialization, positivist science was born as a response to the task of bringing about national unity and modernization: progress based on the control of nature and society was sought; schools, institutes, and scientific societies emerged. Finally, the Revolution once again subordinated scientific development to politics. Each societal change has affected the evolution of the Mexican scientific community.

National Reconstruction

After the first postrevolutionary decade, a period of national reconstruction and of forming modern Mexico began. The Mexican state was restructured and strengthened, and the foundations were laid for the large infrastructural projects that would allow the development of industrialization, the renewal of the educational system, and the ideological unification of the country within a nationalist model.

Within this framework, new institutions of higher learning and research were created; that is to say, the context of modernization and industrial development gave new life to science in Mexico. Although the Revolution had put an end to some flourishing scientific institutions, it created others, such as the Directorate of Geographic and Climatological Studies (1915) and the School of Chemistry (1916); the Astronomical Observatory had been preserved. In 1934 the School of Bacteriology was created at the Universidad Gabino Barreda, which would later become the Universidad Obrera (Workers' University); eventually this school was integrated into the Instituto Politécnico Nacional (National Polytechnic Institute) as the School of Biological Sciences.

Most of the new research institutes were created within the national university. From 1929 to 1973 twelve institutes were integrated into the university, some of which already existed under other names, such as the Institute of Geology and the Institute of Astronomy. These institutes grew quickly after 1960, especially between 1974 and 1982.

Other institutions saw similar growth. Thus, in 1960, the Center for Research and Advanced Studies of the National Polytechnic Institute was founded as a center for graduate studies in mathematics, physics, biology, and electronics. All of its students received scholarships. In 1961, the National Polytechnic Institute began its graduate programs in physics and mathematics, and schools of science were founded in Puebla, San Luis Potosí, Monterrey, Veracruz, and Michoacán. Probably the greatest step forward was the creation of full-time research positions. In this manner the professional nature of scientific research, as an independent field, was recognized (De Gortari 1979, 358).

We can say that by about 1960 science had been institutionalized in Mexico and was accepted by Mexican society as a legitimate endeavor. Although this institutionalization came late, it was not unimportant (in Europe this had occurred in the eighteenth and nineteenth centuries [Ben-David 1971]). In 1969 the Academy for Scientific Research was established, and in 1971 the National Council of Science and Technology (CONACYT) was founded; with the latter, scientific endeavors were acknowledged to be important and were promoted directly by the state.

Nevertheless, Mexico's scientific community has grown very slowly. The country has lacked training institutions, and the process of training scientists has been slow and erratic. In Tables 5 and 6 we can observe that the doctoral level—that is, the research level per se—has grown the slowest. In 1979 there was a total of 2,983 researchers at UNAM, whereas in 1990 there were only 2,793, which represented 90 percent of the total number in Mexico. In 1974, 266 new doctors graduated, while

Table 5. Number of scientific researchers at UNAM, by field, 1979

	All Personnel	Masters	Doctors
Agronomy	972	278	122
Astronomy	69	10	23
Biology	424	84	123
Marine science	152	35	36
Physics	390	89	191
Geophysics	95	26	21
Geography	44	22	10
Geology	340	16	23
Mathematics	287	77	124
Chemistry	210	45	93
TOTAL	2,983	682	766

SOURCE: UNAM 1980, 3.

in 1989 only 204 new doctors graduated (see Table 6), despite the intense effort made to provide official support (CONACYT-SEP 1991). This reflects the precarious state of the country's human resources: internationally a doctorate is, almost without exception, necessary for one to conduct scientific research. In addition, according to other statistics in 1974, 40 percent of the researchers worked less than full time. Consequently, from 1973 to 1974 scientific productivity is low: for every 16.6 researchers, one work was published that year, compared to one work per 4.8 researchers in the United States and one work per 2.3 researchers in Israel (Schoijet 1979, 1–2).

The Latin American University

Before describing specifically the National Autonomous University of Mexico within the context of scientific development that concerns us here, we need to point out the special characteristics that are distinctive to Latin American universities in general. Ben-David (1962, 1968) and Van de Graff et al. (1978) have shown the close relationship that exists between the social structure of a country and the development of its

Table 6. Number of graduates of graduate programs, by level and field, 1984–1989

	1984	1985	1986	1987	1988	1989
Total	6,634	7,047	6,896	7,869	9,916	11,159
Exact and natural sciences	268	390	324	561	382	347
Farming technologies and sciences	192	217	245	340	250	377
Engineering technologies and sciences	864	1,018	862	1,227	1,033	836
Health technologies and sciences	1,813	1,913	1,896	2,027	4,503	5,286
Social sciences and humanities	3,497	3,509	3,569	3,714	3,748	4,313
Total by area of specialization	2,749	2,793	3,036	2,939	5,553	6,554
Exact and natural sciences	25	18	11	69	75	26
Farming technologies and sciences	19	42	72	47	63	43
Engineering technologies and sciences	195	239	218	226	270	131
Health technologies and sciences	1,535	1,622	1,572	1,657	4,133	4,976
Social sciences and humanities	975	872	1,163	940	1,012	1,378
Total for master's program	3,640	4,077	3,704	4,758	4,185	4,401
Exact and natural sciences	231	343	285	448	280	296
Farming technologies and sciences	170	173	164	290	184	328
Engineering technologies and sciences	669	776	642	994	760	702
Health technologies and sciences	268	270	319	340	338	262
Social sciences and humanities	2,302	2,515	2,294	2,686	2,623	2,813
Total for doctoral program	245	177	156	172	178	204
Exact and natural sciences	12	29	28	44	27	25
Farming technologies and sciences	3	2	9	3	3	6
Engineering technologies and sciences	0	3	2	7	3	3
Health technologies and sciences	10	21	5	30	32	48
Social sciences and humanities	220	122	112	88	113	122

SOURCE: CONACYT-SEP 1991, 30.

universities. In turn, the development of universities is closely related to scientific development.

After Latin America's independence from Spain, the region's universities experienced many vicissitudes and many attempts at reorganization. The Napoleonic model, organized into schools or faculties, was finally adopted. The main purpose of this scheme was to train liberal professionals: lawyers, doctors, and engineers (Ribeiro 1971, 67–69, 134–38; Silva Michelena and Sonntag 1980, 17–23; Scott 1968; Otto 1982). In European universities, research organizations (institutes, laboratories) were soon added and had certain links to teaching. Research in Latin American universities, however, has been a much more recent phenomenon (Fuenzalida 1971).

In Latin American national universities, teaching has traditionally predominated over research. Each school or faculty is practically autonomous in academic matters. Each career organizes the curricular content of its courses, and students are not allowed to enroll in a course given in another faculty. The teaching is done mainly by liberal professionals who work outside the university and who contribute their experience to teaching. Their pay, usually symbolic, depends on the number of hours taught; hence, their principal motivation is prestige or a desire to serve society. Thus, academic activity has been limited almost exclusively to the transmission of knowledge from one generation to the next, since the university's institutional and normative structure did not encourage the production of knowledge through research. Society recognized the role of the "savant" or "intellectual," who was usually an isolated individual whose independent income allowed him to maintain a fondness for science within the limitations of his economic means and his curiosity. Within this general framework, there were variants in each country. For example, in Mexico as in other Latin American countries some independent scientific research centers were established beginning in the early nineteenth century (Stepan 1976).

Starting in the middle of the twentieth century, important progress has been made in institutionalizing scientific research at the leading Latin American universities; however, this evolution is far from over. In general, the problem of creating a social niche suitable for scientific research has not yet been solved. Even when science is recognized as a career, its social value generally lags behind that of the liberal professions: "Latin American countries have received a negative historical inheritance.... The colonial tradition entails sterility and deficiency in research or in innovation. The model of society from the nineteenth century until 1930 determined the lack of demands, stimuli, motivations, and possibilities for science and technology. Science and technology were incorporated as finished products, on the model of consumer goods" (Kaplan 1981, 4).

Still, the industrialization of Latin America has gradually produced some changes that are reflected in the university structure. The new middle class has gradually displaced traditional elites in national universities, starting with the 1918 Córdoba (Argentina) movement, which spread to nearly all the universities of the region. The old elitist schools became universities for the masses (Trow 1970) or for the use of the middle classes (Silva Michelena and Sonntag 1980, 24–32; Lomnitz 1977, 315–38; Graciarena 1971, 93–142; Tedesco 1971, 143–87). Growth spread from one university to the next, and new fields of study were created everywhere. Eventually research institutes emerged separate from and parallel with the faculties (see Table 7 for the case of Mexico). Just as in France, "the new research fields were not incorporated into the university structure; rather, researchers had to accommodate their fields into the traditional academic frameworks" (Ben-David 1968, 34).[1]

More recently, some faculties have begun to offer graduate studies in various fields; but the corresponding research centers are maintained separate from the teaching, as if research were a distinct activity rather than an integral part of normal academic life. Nevertheless, researchers work in an institutional environment where they are exposed to extracurricular pressures from several directions. In part this is because the large Latin American universities have assumed important sociopolitical roles: they act as means of access to the state bureaucracy for the middle classes, and they serve to relieve pressure on the job market in the bureaucracy and in the professional fields; they act as battlefields for settling certain political differences or as a refuge for dissidents; and they normally provide training for the national system's future political leaders (Lomnitz 1977, 61; Smith 1975; Camp 1976; Graciarena 1971; López Cámara 1971).

This overall situation in Latin America's universities—with differences of degree more than of substance—has translated into difficulties for the development of research. Although science is universal, a small group of

1. Compare the development of linguistics in Germany and France and the effect that the type of institution in which it was developed had on each school (Amsterdamska 1987, 22).

Table 7. Foundation of UNAM's Institutes of Science

Name	Antecedents	Year
Institute of Biology		1929
Institute of Geology	Geological Institute (1888)	1929
Institute of Geography		1934
Institute of Physics	Mathematics Seminary (1932)	1938
Faculty of Sciences		1939
Institute of Chemistry		1941
Institute of Mathematics	Antonio Alzate Society (1942) Mathematics Seminary (1932)	1942
Institute for Biomedical Research		1942
Institute of Geophysics		1949
Institute of Engineering		1956
Institute of Astronomy	National Astronomy Observatory (1929)	1967
Institute for Research on Materials		1967
Center for Nuclear Studies		1967
Instruments Center		1971
Institute for Research in Applied Mathematics and Systems	Center for Research in Applied Mathematics and Systems (1970)	1973
Center for Atmospheric Sciences		1977
Center for Research on Cellular Physiology		1979
Center for Research on Nitrogen Fixation		1980
Center of Genetic Engineering and Biogenetics		1981

SOURCE: Extracted from De Gortari 1979, 357–75.

countries, representing less than one-fourth of the world's population, generates 95 percent of all scientific knowledge (Moravezsik 1980, 160–65). The historical reasons for this are well known, but their perpetuation means that the peripheral countries must still struggle to overcome the lack of a scientific tradition, the lack of a public understanding of the role of science in development, and the lack of an administrative structure encouraging scientists to accomplish their work.

Moreover, the educational system produces few individuals having the knowledge required to pursue a career in research, since it places a higher priority on the number of graduates than on their quality. A significant proportion of potential scientists continues to depend on training abroad; not only does this affect the cost of training, but it encourages the so-called brain drain and prevents the formation of local work groups. Thus, scientific activity in peripheral countries tends to be sporadic and isolated, which affects both the quality of research and its scientific relevance (Moravezsik 1980).

The National Autonomous University of Mexico

As we have seen, the origins of UNAM go back to the sixteenth-century Real y Pontificia Universidad, a colonial institution that for three hundred years molded the minds of the elite of New Spain, characteristically by transmitting conservative religious and political thinking. After independence, and despite a series of closures and reorganizations, no important changes came about either in the university's tendencies or in its operation. It was closed definitively in 1865, and two years later Benito Juárez replaced it with the National Preparatory School, which later led to the creation of a new national university. In the meantime, the professional schools (jurisprudence, medicine, engineering, fine arts, commerce, and others) continued to operate independently of each other and without central coordination (De María y Campos 1981, 26–28; Silva Herzog 1974, 6–11; García Stahl 1975, 87–89).

In 1881, a group of positivist intellectuals that included Justo Sierra, Ezequiel Chávez, and Gabino Barreda submitted a bill for the creation of a new university. Defined as "national" and "independent," though it was to be linked to the executive branch, which would appoint its president and oversee its administration, this university was to be autonomous in its internal organization. Thanks to the efforts of Sierra and Chávez, the National University of Mexico was created in 1910, just months before the outbreak of the Revolution, in response to pressure from young people belonging to the new middle classes (De María y Campos 1980, 52–58).

A speech given by Justo Sierra, the Minister of Public Instruction and Fine Arts, in support of the bill creating the national university, points to the central ideas of the positivists who had promoted it:

> A university is a center in which science is propagated and where one goes to create science. . . . Science is lay . . . and it has no

purpose other than to study phenomena and to arrive . . . at higher laws. . . . All that deviates from this path may be very holy, very good, very desirable, but it is not science; consequently, if science is lay, if universities are to be devoted to the acquisition of scientific truths, they must be . . . lay institutions. . . . This university . . . is a university belonging to the state. . . . For the acquisition of high knowledge, with the guarantee that in [the university] all rights that the Constitution may give it will be . . . respected for carrying out its scientific program. (quoted in Silva Herzog 1974, 16–17)

This university, conceived of as national, liberal, and lay in addition to scientific, was initially made up of the schools of engineering and fine arts (which included architecture) and that of higher studies. In the latter, classes were taught in biology, physics, and chemistry; in 1925, this school became the Faculty of Philosophy, which retained a scientific area. This area separated in 1935 and was divided into two entities: the Faculty of Physical Sciences and Mathematics and the Faculty of Medical and Biological Sciences. In 1939, the Faculty of Sciences was created. Also, in 1915 the School of Arts and Trades had become the Practical School for Mechanical and Electrical Engineers; and in 1922, the School of Industrial Chemistry and the School of Health and Hygiene had been created.

The upheavals in the nation's life were immediately reflected in university life. Thus, in 1912, the first strike broke out in the Faculty of Law (which eventually separated from the university). In 1914, as a result of another strike, the School of Medicine was temporarily closed. Although the university was criticized for its meager participation in the revolutionary struggle, it had become a political arena, going beyond the original intention that it should be a fundamentally scientific and cultural institution. This point should be kept in mind; for it represents a part of university reality, even in less tumultuous times. In addition, this period of agitation and tension in the national university coincided with similar events elsewhere in Latin America and with the rapid transformation of national universities into universities for the masses. In Mexico the period culminated with the 1929 movement and with the new university legislation that gave autonomy to the university. From then on its name has been the National Autonomous University of Mexico, or UNAM. The university has three functions: to teach, to conduct research, and to disseminate culture. The faculties and schools teach; the institutes and centers conduct research; and the institutes and departments disseminate culture (see Appendix 1).

UNAM's statistical data indicate that in 1979, when the present study was being conducted, the student body totaled 294,000, of which 156,000 were undergraduates; the remainder were high school students. The UNAM budget came to 9.559 billion pesos (at the time, 430 million dollars), which surpassed the budgets of several states in the Republic. However, 62 percent of the budget was earmarked for teaching (in the faculties, schools, preparatory schools, and the Colleges of Sciences and Humanities), and only 15.6 percent went to research institutes and centers (both in the sciences and the humanities). There were 1,206 full-time researchers in the area of science.

As for UNAM's growth, Figure 3 provides an idea of the numerical explosion that has occurred, especially since 1970. This growth was managed thanks partly to the creation of the Colleges of Sciences and Humanities (CCH) and the Schools for Professional Studies (ENEP) in other parts of Mexico City.

During the 1970s university life in Mexico was marked by nearly permanent conflict. These conflicts can be interpreted as a struggle, open or surreptitious, among diverse, more or less defined tendencies, which represent different concepts or projects for the university and which can be classified according to two main purposes: the "academic project" and the "political project" (Pérez Correa 1974).

These projects are not merely abstract ideals; rather, they correspond to real functions that the university is performing for the country. This duality between politics and academia produces an internal tension between the university's explicit functions (teaching, research, and disseminating culture) and certain implicit functions that produce internal conflicts demanding the increased attention of university authorities (Pérez Correa 1975, 375). The implicit functions we refer to are all those that arise from external pressure or from the needs of the national system, including the following: to provide social mobility for the middle class; to serve as a regulator for masses of young people who will enter the middle-class job market; to act as a center of social criticism and an escape valve for the expression of dissent in a system that provides few such channels of expression; to serve as a battlefield or stage of political conflict; and to serve as a training ground for the system's future political and technical leaders (Lomnitz 1977).

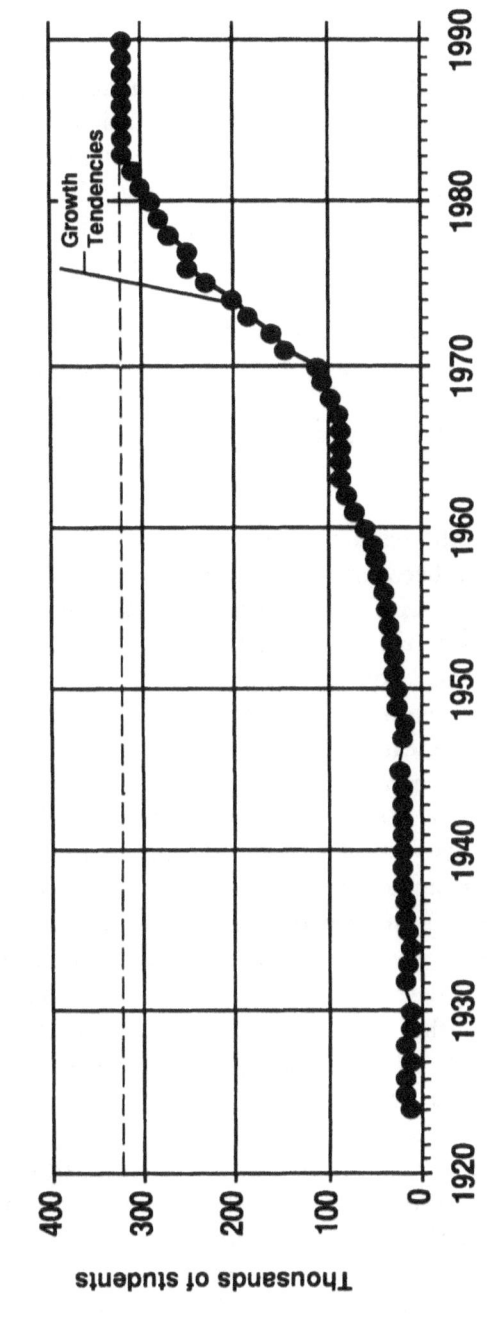

Fig. 3. Total student body of UNAM, 1920–1990

— Real data
• Estimated growth

Source: UNAM 1990, 27.

The internal dynamics of the university should be understood in terms of the diverging and frequently conflictive functions assigned to it by the national system. These functions have become the banners of opposing interests, each of which proposes for the university a different project. The need, for example, to create a solid scientific and technological establishment in Mexico in order to resist the pressure of foreign technology has been explicitly acknowledged (CONACYT 1976). UNAM's scientists and technical experts, however, have seen the frequent political interruptions caused by strikes and work stoppages as an annoyance. For this reason, conflicts have arisen between proponents of the university as the foundation for a national technological and scientific system and those who advocate free expression for all types of internal politics.

At the same time, as the research area (i.e., institutes and centers) grew, conflict also arose between the professional university and the scientific university. One point of conflict relates to the training of scientists, which depends on the professional faculties. The undergraduate program studied in this book represents an attempt to change that situation by creating a program to train scientists that would depend not on a faculty but on a research institute and that would be managed and controlled by career researchers.

To summarize: science in Mexico developed within a historical and social framework particular to a peripheral, dependent society, with a relatively recent and unsustained history of development. Hence, there is still little societal support, affinity, or appreciation for scientists and what they represent to society. The one Mexican institution in which science *has* fundamentally developed is UNAM, which is designed above all to train professionals—that is, users of knowledge, future technicians, and liberal professionals. In addition, the university is obligated to fulfill certain political duties, including the training and selection of state elites.

The university has been transformed into an administrative bureaucracy that handles an important budget. Hence, the scientific community must develop amid political and student pressure, union activity by the administration and the faculty, pressures exerted by the professional sector, and the demands of an ever more complex administrative machinery. Lastly, it faces the difficult task of surviving as a minuscule elite operating within an enormous university for the masses, maintaining and selectively increasing its intellectual and professional demands despite egalitarian pressures that tend to lower its academic level. Throughout this book we shall see how these problems have been faced by a group of scientific researchers in a particular educational context.

2
Origins and Development of the Field of Basic Biomedical Research

In 1940 the Laboratory of Medical and Biological Studies was founded at the National Autonomous University of Mexico in order to make room for a group of Spanish refugees who were students and followers of the Nobel Prize Laureate in Medicine Santiago Ramón y Cajal (1852–1943). Some members of this group had ties with the Rockefeller Foundation and were able to obtain a grant from it to found in Mexico an institution similar to the Ramón y Cajal Institute, located in Madrid, which was famous throughout the world for its studies in the anatomy and physiology of the nervous system (Nieto 1981, 9–12).[1]

The university made it possible for the new laboratory to be established under its direct sponsorship, and it appointed a Mexican scientist, Dr. Ignacio González Guzmán, as the first director (1940–1965). The laboratory initially had three departments: Neuroanatomy, Cytology, and Physiology. Studies results were published for the most part in the laboratory's *Boletín del Laboratorio de Estudios Médicos y Biológicos*.

---

1. See Cueto 1990 on the role the Rockefeller Foundation had on the development of biomedical sciences in Latin America.

In 1956, the laboratory, which had become the Institute of Biomedical Research, obtained its own building and eight full-time research positions. The Institute's budget grew, allowing valuable equipment to be bought and a good library to be formed (Nieto 1981, 9–12). Figure 4 shows the Institute's growth.

In 1965, the Institute of Biomedical Research changed course with the election of Dr. Guillermo Soberón as director. Dr. Soberón had a different background from that of the initial group members: he was a biochemist, he had received his Ph.D. from an American university, and he had formed a group of researchers at Mexico City's Nutrition Hospital. Although they were all trained doctors, his researchers possessed a basic scientific orientation and represented a more recent scientific school, based on the important developments in biochemistry and molecular biology made after World War II. These new concerns represented an orientation fundamentally different from the traditional European school (from which the Institute's founders had come), since they showed the influence of the recent emphasis in English-speaking countries on the study of the fundamental physical and chemical processes within the living cell.[2]

During this period the Institute's structure gradually changed until it had the departments and research areas it has now: Neurobiology, Physiology, Molecular Biology, Developmental Biology, Biophysics, Biomathematics and Immunology. In five years the Institute was transformed and modernized, with the new Department of Molecular Biology standing out.

In 1966, there were 15 research groups; by 1981 this number had risen to 46. The number of annual publications rose continually, from 42 in 1961 to 130 in 1981 (see Figure 4). In 1981, the Institute had 65 full-time researchers, 65 full-time academic technicians, and 160 students (Willms 1981, 16). It had established important links with other Mexican institutions—hospitals, research laboratories, and teaching centers outside Mexico City—both inside and outside the university.

In 1971 the director, Dr. Jaime Mora, encouraged an internal debate regarding the academic goals of biomedical research in Mexico and requested that a study be done on the Institute in order to formulate a true questioning of its structures, hierarchies, and decisionmaking

---

2. For a definition of "schools of thought," see Amsterdamska 1987, 9: "A school of thought . . . is a group of scholars or scientists united in their common divergence, both cognitive and social, from other schools in their discipline or specialty as a whole."

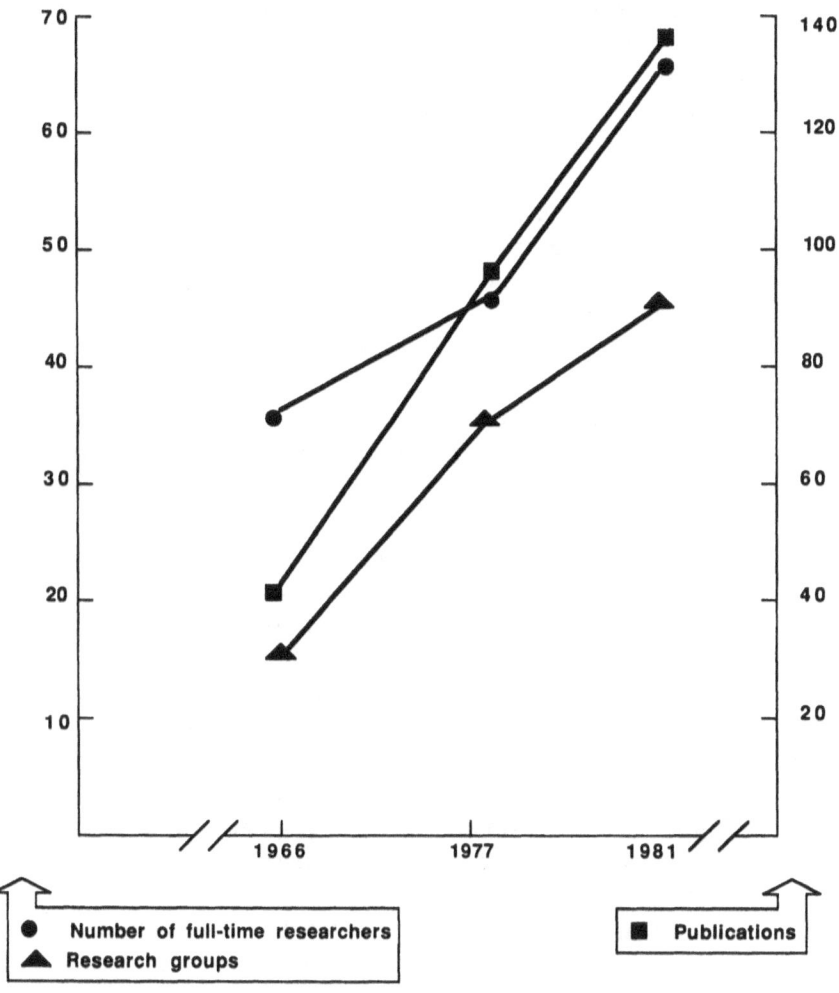

Fig. 4. Number of groups, researchers, and publications at the Institute of Biomedical Research, 1966–1981

SOURCE: IIBM 1981, 16.

processes. This diagnostic study of the internal situation (Lomnitz 1972; Lomnitz 1976) made it possible to observe closely, for a period of five years, how the new undergraduate program in basic biomedical research had originated. The initial questioning was related essentially to the manner in which science is conducted in Mexico and to the relevance of the work for the nation and for world science. This questioning raised other questions, dealing with research policies, with processes of communication among faculty members, with the hierarchies that were established among researchers and, finally, with the whole training process, the product of which was the researchers who made up the Institute's academic staff.

The internal questioning that was produced in the Institute of Biomedical Research from 1971 to 1974 revealed the existence of a series of problems related to the process of training researchers (Castañeda 1974; Castañeda et al. 1975; Lomnitz 1972). The main problems may be summarized as follows:

1. The personal decision to pursue a career in research was the result of random processes. The future researchers were students in professional careers, such as medicine, who had discovered the path to research fortuitously. Their professional training had not provided them with the background necessary for scientific work; hence, it was essential to devote considerable time to them after they had finished their undergraduate studies.
2. Training of researchers was incomplete and late. The experience of having studied for a traditional undergraduate degree limited future scientists and hindered their analytical ability, their basic scientific culture, and their creativity.
3. After the personal decision to pursue research was made, the choice of a scientific field was also random. It depended on the chance of open positions and on fortuitous contacts with advisers or on casual information passed on by classmates; it was not the product of a rational decision.
4. The advisory system was efficient in training scientists; still, it had important drawbacks, owing to the intensely personal and exclusive nature of adviser–student interaction, which seriously limited the student's range of interests. This range of interests was formed under the influence of a single person: the head of the laboratory where the student had fortuitously arrived.

In addition, the type of science that was being produced was questioned. It was felt that, in general, research was neither groundbreaking nor particularly relevant to the needs of the country. This was attributed to a series of factors: an excessive dependence on the formative stage abroad; the organization of science in Mexico; the deficiencies in recruiting and training researchers; the manner of obtaining promotions; the division of labor; the division of work groups into hierarchies; the lack of a mechanism for cooperation and communication; and the absence of critical mass, among other problems.

The Origins of the New Undergraduate Program

It is important to stress the importance of the questions raised as well as the importance of the spirit of self-criticism that was emerging in the Institute at that time in order to understand the primacy that was later given to certain formative aspects—such as early initiation into research, the student's independence from imposed criteria, and creativity—in the new undergraduate program.

The members of the Institute who were most deeply involved in this exercise of self-evaluation and criticism came to think it necessary to produce a new type of researcher, one who would not only have a better scientific training but who would know how to face creatively the opportunities and limitations of the Mexican reality. It was concluded that this was not only necessary but possible and that it was the duty of the Institute itself to carry out this effort, since an intensive and selective undergraduate research program, administrated by researchers themselves, was being proposed, and such a program could not be carried out in huge faculties devoted to training professionals who use information rather than to training the persons who produce it.

The program that was eventually developed was designed by the Institute's researchers, the idea being that they themselves would teach in their own laboratories. The classroom was not to be separate from the reality of research; each student would pursue his or her studies with an aim to pose, design, and solve scientific problems in an active research environment.

The scientists who founded the new undergraduate program had very clear ideas on the type of researcher they wanted to produce. They felt that the prevailing conditions in Mexico would make necessary a nearly total immersion and resocialization of the student. These ideas were consolidated during the formalization of the program and during the corresponding interaction with university authorities.

As for the characteristics of an effective researcher training program, it was concluded that the most urgent need was to create an undergraduate-level program; that is, one with a recruitment age of eighteen years or younger. It was felt that the conditions peculiar to Mexico required an earlier and more intensive scientific training, owing to the deficiencies of secondary education and the need to offset cultural and environmental factors that discourage a scientific vocation. At the same time, this strategy would facilitate educational work by obviating the need to struggle against deficiencies caused by professional schools and would allow teachers to work with students who are less tired, fresher, more enthusiastic.

Moreover, the possibility of using a more open, more critical teaching method was considered. It would be of no use to recruit future scientists at an early age if they were to receive the dominating influence of just one adviser. Hence, it was necessary to design a teaching system open to various advisory influences and to stimulate intellectual independence.

The spirit that led to the creation of the undergraduate program may be summarized in the following words from the Institute's director: "We have begun working with ideas that are not our own, with experiments and data that did not represent an original problem. . . . We followed them to see if we would arrive at an idea. We have to replace that focus with the search for good ideas. The ideas on which we work are not always good ones, nor important, nor ours. We need a change of attitude and a niche for the researcher from these countries, which cannot compete with the large [countries], and a [niche] where our creativity can be expressed. This consists of looking for good ideas, working in groups, and finding relevant problems" (quoted in Lomnitz 1972, 33–34). At the same time, the initiators of the new program had expressed their nationalist concerns. Through self-criticism of their own scientific training, they had realized that graduate studies abroad were just as subject to chance as they had been in Mexico. In the great majority of cases, the choice of one's university and advisers had resulted from a series of coincidences, which had determined one's field of specialization and central topic of scientific interest, not only for oneself but, eventually, for future students. Many such topics had become outdated and were no longer on the cutting edge at the foreign universities where the Mexican student had acquired them secondhand. The Ph.D. dissertation, instead of attacking a front-line issue, often represented a secondary operation to finish off a topic, a way in which to tie up loose ends (Carvajal and Lomnitz 1981, 90–98).

By transferring to Mexico a research topic in which the fundamental advances had already been made by others, the dependent nature of national research was reinforced, and the possibilities of making original contributions having a true international impact were reduced. Moreover, to the extent that a topic had already been studied abroad, even more important, more elaborate resources were required in order to obtain scientifically relevant findings. Such resources did not generally exist in Mexico. Young researchers, having recently received a Ph.D. abroad, frequently found that they lacked the means to continue their research. They would have to change topics or engage in a frustrating struggle against the milieu.[3]

Not only were many of the topics imported from abroad not on the cutting edge, but they did not correspond to local reality or possess direct relevance to the country's needs; they came from another reality, removed from Mexico's. Moreover, numerous topics of immediate importance for Mexico lacked salience at the scientific center and, therefore, were not studied in Mexico either. It was thought that an effective program for training Mexican scientists would simultaneously solve the problem of selecting relevant topics by eliminating the excessive dependence on foreign advice, by more selectively awarding graduate and postdoctoral fellowships. To summarize, the two fundamental objectives for the new program were 1) to form good scientists and 2) to form scientists who were focused on problems relevant to the national reality. Both objectives were reflected in the program and in the socialization goals for the new major.

The tentative guidelines for the undergraduate program covered most of these considerations (see Castañeda 1974), but the first attempts to put them into practice faced numerous obstacles, both in the academic as well as the administrative sphere. Initially a curriculum was developed and submitted for consideration by three faculties: Medicine, Chemistry,

---

3. See Rodríguez Sala de Gómez Gil and Chavero González 1982, 74–82 on the problems faced by scholarship recipients upon returning from abroad.

and Science. None of the faculties accepted the program that had been submitted, which meant in practice that a student who followed the course of study in the proposed form would not be able to obtain a formal diploma. We need to remember that institutes are not authorized to grant degrees, since their specific function consists in conducting research. Researchers can and should teach, though not in the institute itself but in a faculty that hires them for that purpose.

This problem was circumvented thanks to the Colleges of Sciences and Humanities, or CCH, a new high school and teaching-program system authorized to issue university degrees. The new undergraduate program was administratively integrated with CCH, which accepted without objection the curriculum that had been submitted. In practice, the Institute of Biomedical Research administers the undergraduate program with complete freedom. CCH registers students, maintains statistical information, monitors academic output, and grants degrees.

Still, the bureaucratic process of obtaining approval for a new bachelor of science program and getting past the obstacles of commissions, councils, and other administrative requirements took nearly two years. Finally, the University Council granted its approval on October 4, 1973, when Dr. Guillermo Soberón was university president. At UNAM, this was the first time that a basic alternative to the traditional undergraduate programs offered by the faculties had been proposed and approved. In addition, it was the first time that a group of institute researchers was able to impose an ad hoc undergraduate teaching program, academically controlled by them and not by career professionals.

As for difficulties of an academic nature, it had been decided that all classes would be taught in the laboratory, using an interactive method. The idea of traditional class lectures (*cátedras*) as a means of transmitting knowledge was rejected out of hand, and it was proposed that teaching be based on problem solving. The authors of the undergraduate program in basic biomedical research had no experience with this method, however, and so the practical problems implied by an interactive education in scientific research were not anticipated. If it is hoped, for example, that the student will acquire a sufficiently broad scientific foundation by conducting experiments, then it is necessary to have sufficient time and a sufficiently varied number of experiments to cover the subject and allow students to attain the level of knowledge that will finally enable them to conduct independent research. In the following chapter, we shall see how the teachers dealt with this academic-pedagogic problem.

Internal Structure of the Undergraduate Program

The undergraduate program in basic biomedical research is under the supervision of the Academic Unit of Professional and Graduate Cycles of the Colleges of Sciences and Humanities, which "establishes the general normative aspects through its Technical Council and Directive Committee" (IIBM n.d., 5). It is coordinated by the Internal Council, which oversees the program's policies and development. The council consists of the following: the undergraduate coordinator; the director of the Institute of Biomedical Research, who supervises the academic project; two teacher representatives; and two student representatives. Initially, the council members were designated by the Institute's director; later they were elected by undergraduate teachers. The duties of the council are "to approve the academic plans and programs submitted by teachers for fulfilling their work during each school year; to study, discuss, and make recommendations on the activities carried out within the degree program" (IIBM n.d., 5).

In addition, a degree coordinator "organizes and supervises academic activities and programs as well as the Pedagogic Research Area." The coordinator maintains administrative and academic contact with the directorate of the Academic Unit and the Institute of Biomedical Research (IIBM 1981, 5). The degree coordinator, in accordance with the statutes, is named by the CCH coordinator; however, as there were no bylaws until 1974, the coordinator was named directly by the director of the Institute.

The Internal Council selects its staff according to area of specialization and tries to ensure that this is in line with the objectives for the semester. The core of the faculty is made up of advanced researchers, most of whom have master's or doctoral-level studies and many years of experience in the field of research (see "Academic Traits of the Teachers," below). The program was originally housed in the Institute of Biomedical Research, which provided classrooms, study areas, and two laboratories. Students had unrestricted access to material and to the laboratories as well as to the library, which was nearby these facilities. The location of the students allowed them to have considerable contact with their advisers as well as with the Institute's other teachers. In addition, since the practical work would not be restricted to a student's own laboratory, he or she could learn about the Institute's activities. This is in contrast to other UNAM careers in which students not only lack their own laboratories but are assigned to classrooms whose location discourages students and teachers from having closer contact.

The undergraduate program's curriculum was developed in such a way that it could be completed in four years (see Appendix 1). Moreover, the study, discussion, and experimentation activities that made up the program made it essential for students to devote themselves exclusively to their studies. The program was broken down into two basic parts. The first, which covered the first two years, was formative: the foundations were established regarding knowledge and how to work on basic research. In the second part, students dealt with problems of application in the fields of medicine and biotechnology.

The Development of the Undergraduate Program, 1974–1980

The undergraduate program was conceived to admit a small, very select group of students who would receive intensive, individualized training. Naturally, recruitment was a primary concern in this type of educational concept.

The requirements for students to be admitted into the program were as follows:

1. They had to comply with the requirements stipulated in UNAM's General Registration Rules; that is, they must have completed high school with a minimum grade-point average of 7 (on a scale of 10) and, in the case of students from private schools, must have taken the admission test.[4]

---

4. Students of high schools that belong to UNAM or CCH are automatically admitted into UNAM's undergraduate schools so long as they achieve the minimum required grade-point average.

2. They must have passed the selection test developed by the Teaching Committee, the purpose of which was to evaluate students' problem-solving ability more than their knowledge. It covered four areas: mathematics, chemistry, physics, and biology.
3. They must have been interviewed by the Undergraduate Selection Committee, made up by teachers from the program and two psychologists. The objective was to get to know the students personally and to ascertain their interest and motivation. In addition, psychological tests were given to applicants to reveal traits considered necessary for scientific work.
4. They must have attended formal talks on the degree program. One or more talks were given on the characteristics of the research program and on the work of science. It was stressed that rewards were scarce and difficult to obtain; the long duration of the degree program was discussed, as were the dedication it required and the few financial incentives offered. The purpose of this requirement was to create another filter in selecting students, so that those who lacked motivation would be discouraged or would be attracted by the type of values and satisfactions the degree offers.
5. They must have taken an introductory course. Students must have spent from one to two months in such a course, the purpose of which, besides serving as an additional criterion of selection, was to familiarize them with the laboratory work carried out at the Institute and to give them an opportunity to see the type of activity to which they would be devoted. In fact, the degree program began, strictly speaking, with this course, which was ultimately the most important selection mechanism. For some six to eight weeks, students went to various laboratories where they worked on technical problems, which they tried to solve. Each teacher assigned different activities to the students. Generally, they requested that students ask questions, propose solutions or alternatives to solve a problem, or learn a simple technique and work on it. The teachers elicited discussions in which they paid attention to student participation and interest and the degree to which they enjoyed the work. The teachers observed the students constantly and commented on them.

The criteria used to evaluate the students can be reduced to three: *intellectual ability,* evaluated through participation in discussions; *mo-*

tivation, evaluated according to punctuality and how much interest was shown in the laboratory work or in the class discussions; *inquisitiveness or curiosity,* defined as "coming up with several ideas or questions on a problem and being willing to look at it in different ways." Teachers gave great importance to the quality and number of questions asked by students and asked them to propose alternatives or solutions to problems. One teacher said: "I suggested that [X] not be accepted because he had no ability; his participation was nil and, although he had good oral skills, the content of what he said was without value. . . . I conducted discussions on small projects. . . . The evaluation of students on academic goals and on attitudes took place every day" (I/1st).[5]

Rather than to impart knowledge, the introductory course was designed to present a condensed experience of research work and, through this, to learn about the candidates and allow them to familiarize themselves with the type of work to which they intended to devote themselves. Of the ten candidates in the introductory course for the first cohort, four were accepted and three were placed on probation; in the fourth cohort, of the twenty who took the introductory course, six were accepted and two were put on probation. In both instances, the students put on probation dropped out in the first semester.

All ten of the students selected for the two cohorts were between eighteen and twenty years old when they began their undergraduate studies. They exhibited above-average intelligence (IQs of 110–122). Most of the students tended to use a functional or concrete style of thinking, rather than a conceptual-abstract one. In the intelligence as well as in the personality tests (Raymond B. Catell's sixteen personality factors), most of the students reflected traits of an obsessive nature, aggressiveness, independence, self-assuredness, and imagination. (For more details on the results of the psychological tests, see Appendix 3.)

Most of the students selected for the first cohort (eight of the ten) came from private schools, and there was only one whose parents are not professionals. The parents' professions are shown in Table 8. It should be noted that both of the candidates for the third cohort who were not chosen for the program were of lower-middle-class origin (the son of a worker and the son of a low-level government office worker). The same thing occurred with CCH students whom the Institute attempted to

---

5. This form of notation will be used throughout to indicate the cohort and semester of the students whose opinions have been gathered herein.

Table 8. Students' parents, by occupation, first cohort

Occupation	No.
Fathers	
Government worker (nonprofessional)	1
Doctor	2
Researcher (biomedical)	1
Accountant	2
Chemist	1
Merchant (pharmacy owner)	1
Lawyer	1
Executive	1
Mothers	
Nurse	1
Housewife	7
Teacher	1
Professional (chemist)	1

recruit to the program. These students, for the most part, come from the lower middle classes and have aspirations to rise socially; for them, the research degree is an unknown and is unattractive as a means of social mobility. This was seen clearly in a 1977 survey of 2,000 first-year CCH students, among whom fewer than 1 percent stated they were interested in pursuing a degree in research, whereas nearly all showed an interest in traditional professional degrees (Acosta et al. 1981).

The social class of the candidates had a certain influence on the final selection. Those who came from homes in which there was more information on careers in research and greater contact with scientific fields tended to understand more fully the spirit and ideology of science and showed greater interest in it.[6] It is important to note that all ten students who passed all the tests and who were admitted unconditionally completed their degree, and most have gone on to the Institute's master's and doctoral programs.

6. The same is true of some ethnic groups in Mexico, such as Jews, who have developed a special respect for knowledge. By contrast, a study on Mexican entrepreneurs (Lomnitz and Pérez Lizaur 1987) found that the children of entrepreneurs rarely become interested in pursuing an academic career, even when they have the wherewithal to obtain the best possible education. It seems that their value structure and consequent world vision are important factors in their selection of a career. One exception to this can be found in Jews, who, although they may come from families of merchants and entrepreneurs, choose scientific careers, possibly because of their cultural and religious values.

Work Routine

Allotment of Space

The undergraduate program was developed in the facilities of the Institute of Biomedical Research. A special area on the Institute's ground floor, where students spent their entire days, was assigned to the program. Initially, students had little contact with the Institute's researchers and staff, who worked in their own laboratories on the upper floors. Starting in the second year, however, the students were gradually incorporated into the normal work of their advisers' laboratories, which became the customary place of work.

In the third year, the students of the first cohort left the Institute to attend classes at the General Hospital, the Military Hospital, and the Chiapas regional center (CIES). The purpose of these classes away from the Institute was to familiarize the students with different scientific environments, with different issues, and specifically with issues more closely related to the national reality. But the students missed the Institute, with its good laboratories and careful work methods; they had problems adapting and felt they were wasting their time. As a result of their complaints, the objective of the visits was reexamined and it was decided that the students would be sent only to institutes within UNAM. From then on, the length of visits away from the Institute has been reduced to a minimum, although students in the fourth year have been allowed to choose advisers from other institutions and to spend time away from the Institute, as the third cohort did in the Nutrition Hospital.

Allotment of Time

From the outset, students were subjected to an intensive work routine. For the first year the study schedule was from 8:00 A.M. to 7:00 P.M., with a break from 2:00 P.M. to 4:00 P.M.

During the first semester, students took twelve hours per week of classes in theory, eight hours of seminars, and six hours of laboratory as well as ten hours divided among the remaining classes and for individual study and laboratory practice, depending on the assignments given to each person. Students remained in the workplace all day, which caused many complaints, since they were accustomed to spending half their day at home: "I accomplish less by being here all day. I prefer to go home after 2:00 P.M. . . . I am fed up with being here so many hours; I like to study however and whenever I want" (I/1st).

In the second semester, the class load in theory was lighter, and students had more freedom to manage their time according to their needs. There were no mandatory, set schedules and students could leave the Institute at any time; still, it was observed that they preferred to remain, not just until 7:00 P.M. but sometimes as late as 10:00 P.M. They would also go to the Institute on weekends and during vacation periods. This was necessary because of the intense pressure of the academic demands and the type of experimental work they carried out, which included slow operations at irregular intervals. In the first year, students complained a lot about what they felt were excessive demands that required them to sacrifice interests not related to research. Nevertheless, they grew accustomed to working long hours and punctually turning in all work necessary for the development of their experiment, regardless of the time of day.

Beginning with the second year, the schedule was more flexible: except for six to eight hours per week devoted to classes, seminars, and talks the students' time was distributed to suit their own needs. It was the students' responsibility to manage their time and adjust their pace of work so as to meet the goals set down by their teachers. For most, this represented a serious learning problem, and many had to give up their hobbies and other extracurricular activities because of a lack of time.

Nonetheless, after the first year it was observed that the students had acquired new work habits, reflecting greater adaptability and pride in themselves, as well as a positive appreciation of intensive work. When the pace of the work slowed down, as occurred in the fourth semester of the first cohort, students missed the pressure: "We had so much freedom that we worked very little; we did almost nothing. I went back to my old hobbies and I read a lot. At the beginning I thought this was good, but later I was exasperated because I had nothing to do" (I/4th). At this stage, students still needed to be pressured by their teacher in order to maintain an intense work pace, even if they already realized that such a pace is necessary.

By the final year of the program, when many students were already working on their thesis topic, the work pace had been assimilated sufficiently and was no longer questioned. The problem of knowing how to organize oneself and to administer one's time persisted, but by then this was accepted as a definitive and permanent aspect of a researcher's life. It was also observed that the third cohort, which was introduced to experimental projects much earlier, had fewer problems in accepting the intense work schedule.

Academic Traits of the Teachers

The core group of teachers—researchers at the Institute of Biomedical Research who had proposed and planned the new undergraduate program—was made up of six persons: most were biochemists, all had initially been trained as doctors, and five had later received Ph.D.s abroad. Nevertheless, most of the first-semester classes were taught by other teachers, who were assigned the basic science subjects, although the core group was in control and made the principal academic decisions for the undergraduate program, through the Internal Council.

The first undergraduate coordinator belonged to this group, which always maintained a very active interest in the degree program and was alert to all the program's problems and vicissitudes. From the outset, career teachers, devoted solely to the undergraduate program, had been hired, and other Institute teachers and their assistants were invited to give teaching and research seminars.

Moreover, at intervals, the teachers-ideologues-founders offered research seminar courses. Most of the Institute, including the graduates, have been involved to a greater or lesser degree in the development of the undergraduate program and in the master's and doctoral programs that were later created, although not all of them always had the same dedication and enthusiasm for teaching that characterized the founding group. (See Table 9.) This was reflected in the teaching results, owing to the crucial importance of the adviser–student relationship. Students saw a clear difference between teachers who were committed to the program and those who were not.

Teachers belonging to the Institute shared similar points of view on the program's objectives and methods. They felt it was necessary to train

Table 9. Profiles of teachers at the Institute of Biomedical Research

Year	Cohort	Founders (Ideologically Involved)	Other Institute Researchers	Assistants
I	1st	2	4	2
	3rd	2	3	2
II	1st	3	3	3
	3rd	2	5	4 or more (group)
III	1st	5	3	0
	3rd	2*	6	several
IV	←─── thesis advisers, several away from Institute ───→			

Year	Cohort	Teachers Hired to Work Full-Time for Undergraduate Program	Away from Institute	Shared Ideological Behavior of Founders	
				Yes	No
I	1st	3	0	5	1
	3rd	3	0	6	2
II	1st	0	0	5	1
	3rd	1	1	5	2
III	1st	0	6	3	8
	3rd	0	2*	6	2
IV	←─── thesis advisers, several away from Institute ───→				

SOURCE: Authors' research.
*Includes teachers on leave from the undergraduate program.

researchers through "nontraditional" teaching methods: problem solving, discussion, laboratory work, encouragement of student rebelliousness, questioning, and the primacy of "formation" over "information" were stressed. The emotional charge of these opinions produced a mystique that was shared by everyone except a small minority of teachers from outside the Institute.

Of the nine teachers who taught the first cohort during the first year of the program, three had been hired specifically to teach basic subjects and to provide general tutoring: one had a bachelor of science degree in biology, and two had master's degrees. Two other teachers were physicians who had studied in Mexico and who had pursued supplementary studies abroad; the four remaining teachers had, in addition, doctorates in science (two had studied in Mexico and two had studied abroad), and they had done postdoctoral work in the United States. Out of fourteen teachers, thirteen had doctorates in science (nine in Mexico and four abroad), and all had done postdoctoral work abroad. The exception was one teacher hired who had only a bachelor's degree. This high proportion of teachers having doctorates is unusual at the undergraduate level in Mexican universities.

It is significant that the second year of the program, which was directed almost exclusively by Institute staff members having doctorates in science, was generally considered by students to be the most important and satisfying, both because they began to work exclusively in the laboratories and because of the efficiency with which the principles of scientific research were conveyed and understood. In the third year, when the first cohort went to study in the Military Hospital and the General Hospital, the students encountered teachers who were medical doctors but who had done no postdoctoral work in scientific research. Not only did the students notice the change in methodology, but they had very strong adverse reactions. They complained about three things: the new teachers' pedagogical ideology, which was more authoritarian and intolerant of critical discussion; the focus of clinical research; and the impersonal treatment they received. As a consequence of these complaints, successive classes' visits away from the Institute were more carefully planned and, beginning with the third cohort, courses given by non-UNAM teachers had to meet certain minimum conditions having to do with pedagogy. The same criteria were applied to UNAM teachers who did not belong to the Institute.

The experience gained from this cohort, especially with respect to the process of learning observed by the teachers in the students, caused the former to reformulate the nationalist aspect of their ideology regarding the importance they placed on creating scientists who would be sensitive to the country's needs. The teachers decided to suspend these external courses and to concentrate coursework at the Institute, complementing that work with some closely monitored courses in other centers, taught by teachers who shared the same general teaching ideology and methodology. Moreover, this represented a change in the ideology of the teachers, who discarded their nationalist discourse and redirected their efforts toward training solid researchers (which, ultimately, would produce scientists who could apply their knowledge in solving the country's problems, although the latter goal would cease to be one of the undergraduate program's explicit goals).

From then on, teaching was developed entirely within the Institute of Biomedical Research. Each student chose the laboratory in which he or she would spend the year, under the supervision of the adviser–head of the laboratory, developing a given research project. Only two teachers came from outside the Institute: one was from another UNAM institute, and the other had been one of the Institute's founders but was now working away from UNAM. The results of this program change were judged to be highly satisfactory.

During the fourth and last year of the undergraduate program, there was no formal teaching in classes or seminars; rather, the students focused on developing their theses under the supervision of their respective advisers or laboratory directors. Seven teachers participated. Six were IIBM researchers who belonged to the Department of Biomedicine and were in charge of a laboratory; the other was a researcher in pathology who had previously worked in a high-level institute and who had also been a researcher at the Institute for many years and a teacher of some of the Institute's researchers. This last researcher had three assistants to teach his classes, all of whom were medical pathologists, including one trained researcher.

The First Year

The development of the degree program for the two cohorts analyzed here was slightly different because of the program's initial newness and because of the characteristics of each cohort. The first year of each cohort was marked by uncertainty and tension. It was an arduous learning experience for teachers and students alike.

For the teachers, the first year of the first cohort represented the beginning of the fulfillment of their ideals: the possibility of molding a new type of researcher. Even though the teachers had planned it in detail, the undergraduate program was still somewhat new for them because of its pedagogical focus, the teaching methods required, and the characteristics of the students (young, recently out of high school). For the students, the program represented a new and difficult world of which they had no knowledge from previous cohorts, as they would have had in the traditional degree programs offered in the university's faculties and schools; in addition, they realized they were entering a degree program that was more demanding than the usual ones. The teachers' lack of experience was reflected in the sometimes contradictory messages they conveyed to the students, not to mention the changes in curriculum and method that occurred before the third cohort entered.

Although it had been planned that the first cohort would begin the year working on problem solving, the students actually spent the first semester in theory courses that made up for their lack of knowledge in basic subjects. Not until the second semester did they begin to work on biology problems in the laboratory. They solved simple biology problems and had to find the required information on their own. The students did not know how to look for the required information, which was at times overabundant, at times incomplete; they worked hard, but they did not get results and worried about the teachers' reserved attitude, which they interpreted as one of disapproval.

For the third cohort, on the other hand, basic theoretical knowledge was taught during part of the introductory course, and the students began their first semester by doing research, although two methods were tried: part of the students began working experimentally on a biology problem; the other group first examined information, then began to experiment. The first group (which was called the "theoretical group") was more conflictive. It was difficult for them to work experimentally and at the same time look for information in order to understand a problem; they examined too much information, which served only to confuse them further. Thus, they required more time to conclude their work than the other group, which did a more traditional search of the literature before initiating the experimental work. The entire cohort later devoted the end of the first and second semesters to working experimentally and to attending information-discussion seminars as a complement.

The first year was characterized generally by the teaching of a discipline of work—reading, data searches, laboratory techniques, scientific knowledge, ways of thinking, criticism—rather than by the generation of ideas. Nevertheless, the importance of being creative and innovative was repeatedly stressed.

For all the students, the beginning of the degree program was a drastic change from what they had been accustomed to in high school. It was very difficult for them to adjust to the teachers' new demands and style.

The students agonized over the uncertainty of their ability to become researchers and over their output, since they never felt they produced enough in their teachers' eyes and were also unsure about what to do to gain the teachers' approval. The teachers' demanding attitude (they would press and scold the students frequently) and the numerous mistakes the students made in seminars and in experimental work deepened their discouragement and frustration, but also increased their desire to know what their teachers thought and to think like them.

The pressure of work kept the students overwhelmed. They were required to absorb a lot of information—so much, in fact, that they failed to distinguish between what was essential and what was not; they had so much work that they were unable to divide it or rank it according to importance. Following this, there was another stage in which a drastic change arose. The teachers held the students to very high standards and established general objectives (to research a problem, to solve a question experimentally, to look for information on a topic) without distinguishing between the different steps needed to fulfill the objectives.

The following quotations illustrate the students' level of socialization during the first year:

> I learned to set my own goals . . . not to believe things so readily, because at the beginning I took everything the book said literally, and now I don't; now, when I read an article, I analyze how the author found that out, what he did to demonstrate it; you become more critical in reading and discussing. (III/1st)
>
> In science . . . most of the time you work to know and to figure things out . . . regardless of the immediate application . . . and that's really neat. Research is always good for something . . . and [the findings] are useful even if it's to say that the data are no good. I'm not worried about the possible usefulness of the data. Merely asking yourself questions and looking for the answer is important. The discipline of thinking about everything helps you in life and, in research, makes you think a lot. (I/1st)
>
> I sort of learned to think, to see that what we know we got from reality; that's where knowledge came from, more than from abstract matters. Now I try to see, and to think in relation to the things we see. (III/1st)

After his first semester, one student said:

> Now I'm very skeptical. . . . [N]ot until I understand where an equation comes from or what the point of an article is do I feel satisfied. I accept only what is established. I look for convincing arguments. (III/1st)
>
> What I like about the major is that it makes you aware of what you're studying . . . to know that nothing is absolute, that you shouldn't study things without being sure of what you're studying. . . . [I]n other words, [you should be] critical . . . and now I realize how important that is. (I/2nd)
>
> I have learned to ask a lot of questions and to realize that I have a lot to learn. . . . [Y]ou get the idea that if you're going to do something you have to ask the teacher, other people; you start to get over the fear of asking things. (I/2nd)

The second year of both cohorts was similar in that the students did research, with similar methods and teachers. The students joined their teachers' laboratories during the third semester, and in the fourth semester they worked on research in their own laboratories.

By the end of the second year the students had assimilated a work discipline: they would work in the laboratory for whatever amount of time was necessary; they managed their time better and organized their work. The students felt very involved; they had discarded their other interests; and, when necessary, they even went to the Institute on the weekends.

> Last weekend I was helping some first-year kids with protein determination and, it's funny, I look at the test tubes and I see which ones I can use to distinguish a given range. I knew what had happened when they went outside the range. I felt very happy, but they didn't; they didn't know and they felt very unsure, just as I had in the first and even in the second semesters. So I feel more sure of the techniques, even of pipetting; they're very careless and they don't do it right, and I feel really at ease, really great. (I/3rd)
>
> I'm very much at ease in the laboratory; you feel really good about knowing how to do things; now that we have experience, we know how to handle ourselves in the laboratory, and we are more knowledgeable. I'm more comfortable. . . . I think I'm getting into this more. (I/3rd)

By the second year, then, students began to anticipate the course of an experiment and to detect mistakes, which they were sometimes able to correct. They had learned the principal forms of reasoning and how to use the techniques better.

The students had begun to understand something about mental discipline, which they adopted and carried out to an extreme degree. Because of the responses they received from their teachers, the students were aware that they had learned to act out the role of the model researcher, and this reassured them; they felt dissatisfied with what they were doing and with the limits of the degree program. On the other hand, they also felt that they were already researchers; they spoke of themselves as such. One student said: "As a researcher you get used to reading, discussing, and this reflects upon other activities" (I/3rd). The students were happy because they felt accepted by the researchers.

At this stage, the end of the second year, a change occurred with respect to the image the students had of researchers. They found that image difficult to put into words, and they had trouble talking about what a researcher is and what a researcher should or should not be: "Hmm . . . now it's hard to say what characteristics I think a researcher should have. I can't really put my finger on their characteristics; I sort of see it from the inside and, it's hard, I don't really know" (III/4th). When they made an effort, however, certain aspects of mental discipline and certain emotional aspects stood out, especially in the fourth semester.

The students felt that they identified more with researchers, although they also felt an ambivalence, a fear of being too absorbed by their new role. Thus: "I don't like to get along too well with the Institute's researchers because I don't want to fool myself. The Institute alienates you; everything is science, science. I don't want to lock myself up in the Institute or want my group of friends to be here" (I/4th). This student was the most socialized of his cohort. Starting in the fourth year he joined the Institute fully, pursued his graduate work there, and remained to work. He even formed a relationship with someone from the Institute. His was not the only such case.

3

Experimental Research Work: The Teaching Methodology

In the preceding chapter we saw that the members of the program's founding group decided against the traditional system, which was based on courses taught through lectures ("blackboard classes") and which placed students in a passive role, making them accumulate information rather than stimulating them to look for and solve problems (and thereby encouraging them to assume a critical attitude). Hence, the central pedagogical purpose in the creation of the new undergraduate program was to use a new method that "would structure students' minds so as to increase their scientific creativity." To this end it was decided that students would be introduced to research starting in their first semester in the program; that is, they would learn through laboratory research, problem solving, and discussion with their professors that would stimulate the habits of questioning, imagining, and doubting. Students would not be provided with information directly; rather, they would be encouraged to look for information or to discover techniques as they acquired them.[1] (See Appendix 2 for the principal concepts and techniques that were learned in the four years of the undergraduate program.)

---

1. Collins (1985, 159), analyzing the possibilities of change in science and how to train

Nevertheless, soon after classes began, the professors realized that this new orientation could not be adopted immediately, since the students lacked essential basic information with which to look for or solve problems. There arose an imperative need to teach basic courses in which professors, through discussions and by assigning bibliographic information searches and oral reports on different topics, would attempt to encourage student participation and prevent students from being passive.

Although the professors found it difficult to convey information, they also felt that the failure to do so would prevent students from solving problems. They recognized that "training" presupposed the possession of certain knowledge that is important to acquire or to learn how to acquire. One professor who taught a seminar in the fourth semester said:

> Based on my prior experience with the first cohort, I felt it was my duty to provide them with information. On the previous occasion I had supposed that they knew more and I was too certain that everything [that was] informative was bad. . . . It's a pendulum. . . . [B]efore, teaching was bad because it was too informative, and now we make it good by removing the information. I believe that although a considerable amount of research is experimental and manual, it is naive to think that whatever is informative is superfluous. Productive researchers have been trained through informative teaching, and I doubt that, had the informative [aspect] been removed, science would have advanced more quickly in most cases. (III/4th)

Another professor explained his own ideas by saying that

> in the Institute there are two groups: [first are] those who think that the amount of information is irrelevant, that the important thing is to give a few, though basic, concepts, and that the students should find [the information] on their own; the other group thinks that there is certain information, which is not so

---

scientists, distinguishes between "the algorithmic model of teaching (through formal norms) and "the acculturation model" (acquisition of skills as opposed to formal instruction). According to the author, individual knowledge must be acquired by contact with the relevant community rather than by transferring programs of instruction.

minimal, that [the students] must have and that one must oblige them to obtain—whether [they do so] on their own or whether the professor gives it to them. I felt that I belonged to the first group, and I acted in this manner in the first part of my seminar; but in the second part I became frustrated because I felt that the information was not complete. On the other hand, I cannot identify with the second opinion either. . . . I don't know who is right; both [methods] are probably important, but I couldn't just teach [the students] and let them go through what I went through in the doctorate. . . . I began the course feeling that the important thing was that the students review concepts on their own, even if it was just one [concept], but as the course progressed I realized that if I continued in that manner the amount of information I could give them would be minimal, and I began to help them and to review the theoretical concepts. (I/4th)

Hence a conflict arose between the need for information as the basis for scientific work and the importance that the teachers gave to training and creativity. A solution to this problem was sought by having the students themselves look for the relevant information—selected according to the problem being researched—and search out the assigned readings that they would present and discuss in the seminars; therefore, they had to use the libraries and ask their professors or classmates for references.

The classes/lectures were reduced to a minimum. Students were allowed to present and discuss ideas. Moreover, though they fell short of becoming authoritarian, teachers supervised the students indirectly and closely. Regarding this, one professor said: "He didn't respond to the dialogue because, he said, he didn't know physics . . . and I convinced him that he needed to know it, and he began to participate" (III/1st). Students were continually stimulated to speak, to express their doubts, to criticize, and to question the information they had obtained through the reading assignments: "It is important that they criticize everything they read and hear, that they not simply believe everything. We want the students to develop their imagination through the open discussions" (I/1st).

The students gradually began to discuss and criticize what was presented in class. The professor would encourage a student who had been criticized to explain and defend his or her position. This became a characteristic relationship among the students, and between them and the professors. At times this relationship would become extremely contentious. In the seminars criticism began to prevail, often in a hostile atmosphere.

Learning Manual Skills

The initial importance that was given to generating and discussing ideas led students to underestimate techniques and manual practices used in the laboratory (or "kitchen"). Some professors reacted by strengthening these aspects of their curricula: "Technique is used as a tool. . . . [I]t's laborious, it's time-consuming, it requires no intellectual work, but the quality of the experiments depends on it, which is why this mechanical work should not be spurned, since the results of the experiments hinge not only on what one thinks but on how one carries this out. You have to be skillful, you have to develop technical skills" (I/2nd). (See Appendix 2 for the techniques learned in each semester.)

With respect to the development of technical skills, students worked individually on their experiments. A typical situation can be seen in the following observation:

> 9:00 A.M. Chemistry lab. The professor indicates to the students what experiment they are to carry out and how to carry it out. At the end he answers questions: "How do you measure X?" "How many times do you have to do Y?" The professor explains and leaves. Students work alone, each one on his or her experiment.
>
> 10:50 A.M. The professor comes back and checks what each student is doing; he makes some suggestions and leaves.
>
> 12:00 noon. End of the practice session: everyone finished the experiment.

In this example we can see that techniques are acquired through individual practice, with the professor's occasional advice and supervision. At this stage, learning did not take place through direct imitation.

Problem Solving

At the beginning of the second semester it was possible to work using the teaching method based on solving biology problems. The following example shows the teaching method and the learning process that took place.

Students were assigned a specific problem: to draw blood from a rabbit without killing it; to identify, classify, and quantify the blood cells. Students had to apply a technique, without having a hypothesis to prove. The professor presented the problem without giving details on the bibliographic sources that could be consulted or the techniques that were to be used:

> I assign problems to be solved—let the students look for the needed information themselves. [My purpose in so doing is that] they solve the problem on their own, that they look for the information; to force them to think . . . so that they might feel the anguish of having incomplete information. . . . We want to avoid as much as possible [making them] follow instructions . . . so that they will see what the possibilities are and, within these possibilities, see which are most suitable and most feasible, so that they will know why a technique is used. (I/2nd)

Nevertheless, this professor had to assume a more interventionist attitude because of the problems the students had and because of the lack of time. She pointed out problems in the experimental work. She realized that the students neither looked for information nor assimilated it correctly: "The students repeat information without assimilating it, and they give erroneous information." Hence, she devoted several seminars to reading articles, showing the students different ways of reading.

This case exemplifies the teaching methodology most common in the undergraduate program. Student assignments included looking individually for the required information, reviewing methods and techniques used to carry out the experiment at hand, and going to libraries and questioning researchers. Students would choose a specific technique and conduct the experiment at their own pace, although they would discuss the problems and findings in groups with the teacher. In the words of one student:

This semester there is a new system—a few hours of theory class and enough time for other things. In biology a problem is assigned, and each person has to solve it however he can. In a short time I have learned much more than in class. The first thing I did was to look in books [to find out] what blood is. [The professor] gave me one bibliography and I found another in the library by looking up the titles. I spent three days reading. Then I narrowed it down: how to draw [blood], how to separate the different types of cells, how to find them and identify them. Later, I began to imagine a process of how to draw it, and I filled in the process with techniques taken from books, until I figured it out. I drew the blood by making a puncture in the [rabbit's] ear: that technique was mentioned in books, and I had seen a researcher do it. When I drew the blood I had trouble separating the [different] cell types. I had read and learned the techniques and had chosen the least complicated one, and it worked; it was the easiest one. I made a puncture, but too little came out; it clotted; it was full of dirt. Then I decided to draw it from the ear with a syringe. I made some mistakes. In drawing the blood I had to make three cuts beforehand; when I was cutting I made another mistake because I counted in the wrong squares and I got a different number. I had to repeat it. . . . Finally the experiment worked. Nevertheless, I was not happy because I lost part of the blood, and now I am going to have to do it all over in order to count the white corpuscles. The temperature in the refrigerator varied and the blood froze. This was [owing to] a lack of precautions; I should have adjusted the refrigerator before using it and found out whether they turn off the electricity at night, which they do, although I didn't know it. (I/2nd)

Another student described the same experiment as follows:

Before doing anything else I tried to read up. I looked at what I had handy; I tried to find a book that described the blood cells of a rabbit. First I looked through the department's library; I didn't find [any books]. Then I asked the professor and my sister (a medical student) for information, and they both recommended a certain book on histology. From it I took the data I needed to identify [the cells] under a microscope, and I learned the general techniques for seeing them under a microscope. I wasn't sure that rabbit cells and human cells were the same. I tried to find out by going to [the School of] Veterinary Medicine with a friend, who lent us a book that describes rabbit blood. I needed to read up on the exact techniques [for using] a microscope. I spent three days reading. After three days we began to do the experiment, even though we hadn't finished reading, because the professor rushed us. I started to draw blood from the rabbit by making a cut in a vein in it's ear: I had problems with the dosage of anticoagulant that I used; I didn't know why the blood clotted partially. My classmates used another anticoagulant, and the blood clotted. At times I noticed that when we were in the laboratory we forgot to continue reading, for which reason we had problems because we didn't follow the readings correctly. Finally, the professor thought that it was going to take us a long time to find the mistake, and she helped us. The mistake was that we had not used a fixitive on the smear. We all had made the same mistake, because the book we had looked at had a general sort of explanation and did not mention the fixitive. . . . We helped each other in that if I found something I would tell the others and vice versa; we do not give each other all the information we have. The whole experiment lasted two weeks. (I/2nd)

The experiments were not completed in the expected amount of time. The following account on the same experiment by another student gives an idea of the difficulties that arose:

I punctured the heart in the first experiment. I read my book in order to learn what technique to use and to have fewer problems with clotting. I pierced the heart about seven times. . . . I was scared, but I believed that it was the best method. I was unable to draw the blood and had to make a cut on the ear. Then the blood clotted on us, and I had to draw more. I felt frustrated; I had to puncture the heart again. The rabbit died, and we had to draw the blood by cutting [the rabbit] open straight to the heart with the scalpel. I decided to ask one girl who knew . . . what we were doing wrong. It turns out that we had prepared a coagulant for a larger amount of blood, and that was why it spoiled. We had not read that you had to fix the smear, and it was ruined. Everything was frustrating. All things considered, I learned to do it much better. (I/2nd)

Learning to Read and to Question

A second type of learning had to do with reading critically. In the seminars, special importance was given to discussions of the reading assignments. The research seminar on cellular biology, for instance, was described by the professor in charge of the course as follows: "My class is a discussion seminar.... I follow the method of assigning articles; one person presents [each article], and everyone discusses it. I try to initiate the discussion, and I make them discuss it.... Their participation is not voluntary; I make them talk, even though they don't know much" (I/2nd). This professor criticized the students and pressured them to take a position regarding the work being reviewed.

The difficulties the students had in assimilating and applying information to their work can be seen in the following statement by a professor:

> The students had information that was unclear and crude; they looked for it in many books and extracted paragraphs from each [one]; they began to theorize, and they went so far as to invent mechanisms that did not exist. In addition, they were not objective in their interpretations of what they read, and they would make odd assertions. For example, in obtaining a technique for doing something, they would skip details they had not understood.... They were unaccustomed to teaching themselves, and they were not very methodical; they had poor study methods; they copied everything and failed to extract what was most important; they forgot information quickly. (I/2nd)

At this point, some theory seminars were introduced to fill in gaps in the students' information. Several researchers would participate in the seminars at the same time. These investigators' opinions did not always coincide with that of the current teacher, and the opinions were discussed, which constituted a practical demonstration for the students about how to discuss and present different opinions. The students gradually participated in the discussions and expressed their own ideas.

Learning to Identify and Solve Problems

By the time the third generation of the undergraduate program had entered, most professors had consolidated their views on teaching methods and on the problems of the method that had been chosen. The problem-solving method had been established in the first semester. The manner in which it was applied can be seen in the example below.

The professor presented the problem that was to be studied (mitosis) and divided the students into two groups. The first group spent two weeks reviewing the bibliography pertaining to the experiments, while the second group began working directly in the laboratory, looking for information as it was needed. It is interesting to note that all the students expressed a preference for the first group—the "theory" group—whereas all the members of the "experiment" group were assigned to that group by the teacher; that is, they did not join it voluntarily.

The theory group discussed general biology topics, which the students would take turns presenting orally, while the class would evaluate each student's performance. As the professor put it: "They all had to criticize themselves and be criticized by the others" (III/1st). In the experiment group, students would review information with the sole purpose of experimentally addressing the question of how to obtain chromosomes in blood. The group looked for information on techniques, which they would discuss in seminars with the teachers, who guided them so they would choose the simplest one. This teaching method required a greater organizational effort: "We didn't know where to start, what to do, how we should organize"(III/1st). The group made many mistakes and was obliged to repeat the procedures many times, so that, because the group had not obtained the results after two weeks, the experiments continued during the subsequent period, that of bibliographic review. The members of the group agonized because of the short time they had available to complete their heavy work load, and they were frustrated at not obtaining results in the experiments and not being able to read enough.

The professor had anticipated that they would not be able to obtain results in the given period, and for this reason she had provided plates on which they could observe mitosis. Nevertheless, students were dissatisfied and persisted with the experiments until they had obtained their own results. One student described the experiment in these terms:

> I felt weird at the beginning, maybe because I wanted to be in the theory group. During the first week we were really perplexed; we didn't know where to start, what experiments to do; we didn't know how to get organized. The second week we were very rushed. At the beginning we didn't know what to do; we had to read something, at least the techniques, the formulas for each experiment. . . . The professor suggested some books where we could look up the techniques. . . . She recommended a special book where we found what we needed. . . . Finally, at the end of the first week, she suggested a technique, and we accepted it because it was the simplest one. Each one of us prepared his own culture to obtain more chromosomes, which took us two days. The experiment came out wrong . . . and we had to repeat it several times. I had to do the experiment four times to find the chromosomes, but we went a week past the allotted time. I think it's better to begin by reading, because the theory group had two weeks to study and so, when the experiment began, they had had time to get organized, to learn what to do first, what's easiest, etc. And we've wasted more time; we haven't been able to read as much, and I feel that we have too many things to do and study. Now that I've obtained the chromosomes, I have to repeat the experiments that didn't come out right and I have to study. (III/1st)

And another student said:

> The first week we had to do an experiment that didn't come out right, and we're still doing it. The purpose is to obtain chromosomes, but the partial objectives are to learn how to research and experiment, to see what secondary experiments you have to do in order to carry out the main [experiment]. I liked the method because, although I knew nothing of blood cells, I saw differences; but now, with the chromosomes, I can't get the result. The thing is, I feel that having results without knowledge doesn't

work. But I think that part of the method did work, because over the course of the experiment I asked myself questions and set specific goals concerning what to study, which clarified a lot of things for me. The disadvantage is that experimentation takes a long time. In the theory group, they learned more things, but I think I learned better and more thoroughly because [I worked] based on my own doubts. (III/1st)

Initially, the theory group was also disconcerted. The students had to learn to organize in order to find the correct bibliography, and they selected too much information: "At the beginning we were lost, overwhelmed with work. . . . [W]e felt that we should know everything, when [in fact] what [the professor] is asking for is a minimal amount of information. . . . She wants us to question what we read, but we didn't read to clarify our doubts" (III/1st).

Because of their lack of knowledge and the vast amount of information covered in the reading assignments, students found it difficult to discriminate the essential from the nonessential, and this added to the confusion. Gradually the discussions with the professor served to solve these problems, leading the students to consult the basic books on the topic and to handle information more discriminatingly. In the experimentation period students felt very confident: "We had a clear idea of what we were to do" (III/1st). They used the same technique as the other group, as they had decided to do in discussions with the professor. They also had to repeat their experiment several times until they obtained results. One student described her experience in the theory group in this manner:

> When [it was time for] the theory review, there were many problems because we began to study small details and it didn't occur to us to [look for] a book that covered the topic in more general terms. When we had found one, we thought it was great. We had to look for the books and, on one occasion, the professor rejected a book because, she said, it was very bad. On other occasions we would take the book to her after we had read it. Each fact . . . was discussed to see if what we said was reliable. (III/1st)

Another student in the same group commented:

> [The professor] guided us a lot in the discussions, which took place four times a week, [and] in which we would make reports to her. Now we are conducting experiments. Everyone has to prepare his own protocol, reports, data, etc. We talked with the professor regarding what techniques to choose, out of the two or three that we found. She made the decision herself, based on the fact that it was the simplest technique; [however] I liked a more complicated one, but she rejected it. . . . Now I am doing the culture over because it didn't come out the first time. None of us has gotten it right, and we don't know what didn't work. (III/1st)

In both groups, students were to design their own projects based on some issue that interested them, discuss in the seminars the results that could be obtained, and devise experiments and imagine problems. The professor, while allowing the students to use their imaginations, made them analyze the real possibilities for carrying out their experiments and made them criticize each other's ideas: "We had extremely broad goals, grandiose ideas about doing superexperiments, but the discussions made us see to what degree it was possible to do them" (III/1st).

In the last four weeks of the course, the two groups came together again to form a single group and hold joint discussions. The students were satisfied when they finished the course because they felt they were working with the same information. The professor had initially hoped to make a comparison of the two groups, but she was unable to because the experiment group took two weeks longer on the lab work, and the corresponding information had to be conveyed from one group to the other. She concluded that the experiment group had lacked the time to find information and that the students therefore showed no imagination and were content to carry out the techniques in the laboratory. By contrast, the theory group had a wealth of ideas and information; by the end of the two-week period they were proposing imaginative experiments, whereas in the experiment group the only concern that emerged was over technique. Moreover, the theory group found that the two weeks it had for the experiments were insufficient. At the end of the course, though, both groups were using the same information and showed the same degree of technical skill, as was borne out by the final exam.

In sum, the basic teaching method consisted of two stages: 1) looking for information in an atmosphere of criticism and discussion; 2) designing and carrying out an experiment. Although these two stages were not applied in the same order, the results were similar. The bibliographic research was intended to familiarize students with the use of libraries, to teach them to discriminate between relevant and irrelevant information, and to show them how to read critically. The basic teaching unit was the discussion seminar, in which one or more professors would participate and in which information was required concerning the reading assignments and/or the progress and difficulties in experimental work.

Working in Research

In the second year of the degree program, teaching was the responsibility of the adviser of each student and his or her group. The students focused on looking for problems, which were then formulated as hypotheses, and on designing experiments to test them. They learned to develop protocols, carried out the experiments, and analyzed their results. According to one student: "You might say that this is the first semester in which we began to learn research design; last semester we had to count red corpuscles, but now there are techniques. On the other hand, the idea now is to let your imagination run free; the idea is to create different experiments and see which one we think works best" (I/3rd). The students began to anticipate the course of the experiments and their possible consequences, as well as to detect mistakes more quickly and try to correct them: "We have to know what it is that one expects from the experiment. When it doesn't work out, [we have] to explain why it didn't work out and redesign it" (I/3rd).

The following is a professor's description of a genetics course:

> We began from a low starting point: isolating the material with which we are going to work. This takes time. But the experience is valuable, and it will help us quite a bit for future experiments. The standard techniques are in the books, although they can be made up as well; in this case, we used one that had already been made up. I gave them a clue as to how to find the technique easily. They already knew where to look for formulas; there is a manual

where they can be found. I told them that there is a manual and where to look for it. The students grew the bacteria and made patches; with a sterile toothpick they took the colony and spread it on another box, to disseminate the colony so that it could be used as a sample box.... They already knew this methodology; they had seen it in the introductory course. Later on, they saw that some of the colonies had the growth characteristics of the *Escherichia coli*.... I proposed that they look for different media to identify the cells.... They knew that bacteria did not grow in biliary salts because they had read it.... They proposed a medium, and I proposed a different one to them because [theirs] was very long and uncreative.... I gave them the medium and the formula for how to prepare it, and they were thus able to isolate the bacterium. They had a control [mechanism], an *Escherichia* laboratory bacterium grown in isolation, to compare with the one they were isolating. From that they deduced that the cells that had been isolated by them had a growth pattern that was the same as that of the control cell. [This problem] took five days. They read the description of the bacterium, the methodology for preparing media, in medical microbiology manuals.... The second problem was the isolation of bacterial viruses capable of infecting the cell that they had isolated as a means of genetic transmission among different cells. They had to find a source of bacteriophages capable of infecting [the cell]. Where are we going to look for the source? They thought of all kinds [of sources]: sewers, Xochimilco [the floating gardens, southeast of Mexico City], etc. The students brought some sewer water and centrifuged it. Then we began to discuss how to put the phages into contact with the bacteria. I asked them what alternatives they saw. In discussion, they discarded the liquid-medium alternatives: there was a heterogeneous population; we would not be able to distinguish which had [infected the cell]. Then they thought about the solid medium. But how [would they] make it? I gave them some information: in order to cling together, phages require ions and this must be accomplished in surface water so that [the result] can be seen more clearly. And it came out very well [for them]. They obtained several phages, and we isolated three or four different ones. From these, some were chosen at random and it was decided to separate them, and at that point I again told them how to go about it. They provided many ideas, but

I pointed them to the fastest and simplest way. The strategies they proposed had a high degree of risk and were quite tedious. Giving [them] this technique was more productive. . . . After isolating the phage, we prepared it and made it grow on the intended cell, and they obtained a number of infected and broken cells, which is the effect the virus has on bacteria. (I/4th)

Below is a student's description of the same experiment. It shows an increasing mastery of scientific language and scientific thinking:

The professor proposed a problem to us and asked whether we wanted to do it and whether it seemed interesting. We thought about it and proposed a way in which to do it. We came to the conclusion that in order to isolate phages we would need a fine bacterial strain because phages are specific. We had read in another semester that bacteria can be identified through the phages that lyse or attack them. Then we asked the professor if we were to isolate the strains or if he was going to give them to us. He said he preferred that we isolate them. First we designed, a priori, the manner in which to isolate [the strain] and then we read and we discussed it with him. We obtained the fecal matter. We purified it (we tried to isolate the bacterium, to strain it and centrifuge it, etc.). Then we planted, or sowed, a specific medium. We read what the bacteria of feces were and how to separate them, and then we ran tests to see if they gave positive results or not. The bacteria grew and [were] pure because we had chosen a cell progeny. It took us about two weeks to talk, read, grow. The first experiment was growing the bacterium. It was simple because we were able to plan without reading, and we only had to read specifically on the bacterium, because by then we knew how to isolate and culture a bacterium. Then we had to isolate phages. We had never worked with phages, but we had some knowledge of them. We isolated them from dirty sewer water, taken from an open ditch. We centrifuged the water. Then we inoculated the water in a culture in different amounts. The next day we saw that there were phages, because they make little holes (lysis plates are formed) in the bacteria. We had learned that technique from the professor of the previous semester, in the anthology [he had assigned]. This was done in one day. We had ten little holes, and we had to take each group and put [the bacteria] back in the different boxes of bacteria to separate the different phages and increase their numbers. Phages are very difficult to identify, and we don't have the knowledge to do it. First we obtained a few and increased them by sowing them in another box, and we made a phage suspension. Then we tried to determine whether they are transductant. To do the experiment we needed to know how many phages we had, because we needed twice as many phages as [we had] bacteria. For this, we made dilutions of the solution we had, and in the weakest ones you can count them, in order to extrapolate later to the more concentrated ones. We had problems there because at the beginning it turned out to be too concentrated, and we had to repeat the process. This technique is called phage titration. This was new. We had read; we had an idea; and this is not a difficult technique. The third experiment was in transduction. We put those phages with another strain in a medium with fewer amino acids than necessary. What we saw was whether the bacteria had been identified by the phage. But they hadn't; it appears that phages are not transducers or temperate. We did not yet conclude anything, because we had a problem with the experiment. We made a control with all the supplements required by the bacterium, and it didn't work; so we don't know if the medium is good or not. From among the various control experiments we did one transduction with a laboratory phage that had already been tested, and that experiment turned out nicely. But from the phages that we removed from the water, we still have not obtained transduction, and now we don't know if this was because of poor culturing technique or because they aren't temperate. (I/4th)

In a second seminar, the purpose was to study genetic exchange. This is how one student described the experimental work of this seminar:

The professor gave us two tubes of strains in a liquid medium, and we had to tell which of the two was lysogenic. Each one [of us] took a milliliter for our experiment, and we chose different experiments. I read up on techniques and grew the strains in lines next to the *Algorela coli,* hoping that in the areas where the lines crossed there would be no growth, but the experiment didn't come out right; for it to come out right I would have needed a more sensitive strain, like [the one] my classmates had, but I didn't want to do it [that way] because I thought it would be a repetition. I repeated the experiment using another strain, but with the same growth technique—that of crossing lines—and it came out right. Each one of us did it a little differently. My experiment was a little easier because it required less handling. (I/4th)

Another student said:

I made cell cultures and put different chemical agents in them that are supposed to be able to make the phages free themselves. The experiments didn't come out because those strains are freed only with certain liquids, and we found that out later. My [experiment] didn't come out. (I/4th)

At the end of the second week of the course, the professor discussed the results obtained by the students, analyzed their mistakes, and made recommendations on how to conduct the experiment again and obtain the desired results: "He told each of us important things for doing the experiment right." Finally the different interpretations of the results were discussed, with the professor acting as moderator; he said, "I tried to get the information from *them*" (I/4th).

In the second experiment for this seminar the professor participated more actively. Although the students had to look for the information on their own, the professor discussed the different propositions with them and recommended how to do the experiment. The professor said:

The idea was to study an exchange mechanism—transduction: [this] consists of [genetic] information passing from one bacterium to another through a virus. I asked them a question: "What are the exchange mechanisms?" and I left. They read a lot, even though they already had information; . . . based on that, they made specific proposals for experiments, with protocols. (I/4th)

In a group interview, students described their experiences as follows:

The professor gave us a problem: how to obtain mutants. We had to design the experiment. We read a little; we saw that there are different ways of mutating. We met with him, and each one of us told him the technique that he [or she] had chosen, [each one of] which was different. The professor gave us four different types of mutations, and each of us chose one. He gave us something to do, but he didn't tell us how to do it. We selected ultraviolet light as a mutagenic agent. We have to produce the mutant and tell what part the mutation is in. It's difficult because there are a lot of technical details. (I/4th)

I don't know what the potency of ultraviolet rays is and, in addition, you have to follow many steps to obtain this light; I overlooked a detail and did it wrong, so I'm repeating it. There are so many steps that it takes you about two days, and it's difficult to get organized [or] even to go eat during [the experiment]. Now I'm preparing the material in order to begin the experiment tomorrow; I tried to get organized, I made a protocol. (I/4th)

To conclude this seminar the professor focused on a theoretical analysis of conjugation mechanisms. Owing to time constraints at the end of the semester, he taught classes (discussions) exclusively on theory. The foregoing exemplifies the constant information and integration challenge that the professors faced and how they adapted their methods according to the students' progress, the complexity of the problem at hand, and their own judgment.

Discussion

At all levels of the undergraduate program, learning is promoted through discovery. Among the professors, two schools of thought predominated. One favored the production of ideas above everything else and sought to encourage creativity in such ideas, letting the students' imaginations wander. The second stressed methodological discipline as an essential element for the researcher, seeking new methodologies and techniques to solve a theoretical problem: "Good ideas exist, and in large quantities, but few methodologies can carry them out; . . . having a good idea is all right, but you have to put it into action; otherwise, everything is irrelevant" (I/3rd).

The professors gave much weight to students' dedication to their research and to seeing that they worked on it full-time. They also encouraged methodical, rigorous thinking and meticulousness in manual (laboratory) as well as intellectual work. The students generally carried out their work without the constant supervision of their teachers, who acted as advisers on any question to be discussed. Naturally, each course had a different style, depending on the values and style of the teacher.

The essential principle behind the various teaching methods was that of encouraging the student to play the role of researcher and to go through all the stages of a research project: posing a problem and proposing and designing strategies to solve it; conducting the experiment; analyzing the results; drawing conclusions; and posing new problems. Thus, one teacher stressed experimental work as the starting point for theoretical knowledge: "The idea is to teach them what biology is and how to work experimentally. . . . Faced with an experimental description of biological phenomena, students learn to ask themselves questions and to design experiments based on [those phenomena]. . . . Information per se is not the purpose of the course. . . . [Information] is something secondary that is attained" (I/2nd). In addition, the purpose was to apply other knowledge to biology problems and to show the interrelationship among mathematics, physics, physical chemistry, and biology. Moreover, the professors would develop a list of topics to be covered in the course they taught, identifying the necessary theoretical elements and planning in advance, in conjunction with their research group, the experiments appropriate for covering those topics. Nevertheless, the professors stimulated students to propose experiments themselves, guiding them indirectly, in such a way that sometimes the students would even propose experiments similar to the ones that had been planned: "Before assigning them the experiment we agree on what questions to ask and how to ask them; and during the discussion we tend to make [the discussion] as open as possible, [encouraging the students] to propose whatever comes to mind, and as the discussion advances, useless [ideas] are discarded and the students themselves propose various experiments" (I/3rd).

After conducting an experiment, the students would analyze the underlying theory and would recommend bibliographic reviews to each other. According to one professor:

> There is one big experiment for the course [which], in turn, is divided into sequential stages, going from the simplest to the most complicated, and . . . we have, twice a week, joint seminars of all the [semester's] professors and students, covering a series of topics related to what we want to assign in the form of experiments. . . . We began with a problem from which an experiment was planned [and] from which others were derived. From the observation and comparison of different strains, for example, more experiments emerged, which we stimulated [the students] to propose. . . . In the discussion, we would ask them what characteristics a culture medium should have; one of them said that it requires oxygen or carbon etc. In this way we would guide them and, in special cases, they were given bibliographic information, but always in the light of experimental data. (I/3rd)

In the discussion seminars, different ideas, possible experiments, and their difficulties and results were proposed and analyzed, as were the related theoretical aspects; in addition, new questions and hypotheses were formulated for future experiments.

In sum, the examples described here give us an idea of how the students were taught to be researchers by acting out a researcher's duties. In all the seminars, in one way or another, the need to be creative, methodical, disciplined, and critical was stressed; importance was given to reading and to laboratory techniques as well as to discussion with classmates and professors. The basis of the method was experimentation, reading, and discussion, under the constant supervision of professors. The undergraduate program's training process culminates with the thesis, following the last year of coursework. The students conduct their thesis research for about a year, within the context of work carried out in the laboratory; students must propose a problem, carry out the necessary experiments, draw conclusions, and write their theses. It should be noted that this represents only one stage in the researcher's training, which continues with the master's- and doctorate-level studies.

4

Ideology and Socialization: The Ideal Scientist

The ideological system plays an important role in every socialization process; it is the vehicle that allows the social system to be transmitted and reproduced over time. Having a common ideology provides people with a feeling of solidarity and community and sets them apart from other groups. Althusser (1976, 133–38) has shown how ideology acts as a mechanism for the subjection of a society's individuals, by transforming individuals into subjects who adopt as their own the ideas and the values of the prevailing social system, which in turn gives a mirror-image identity to the subjects.

Scientists constitute a community oriented to the production of knowledge. To become a member of this community, individuals not only need to learn a body of knowledge, techniques, and forms of relating and behaving; they must also internalize an ideology through which they develop a controlling structure, a "superego," which will determine to a large degree their manner of behaving and thinking.

In this chapter we present the ideal model underlying the socialization process, which we will describe in the next chapter. We developed this model as a way of detecting the values, attitudes, and beliefs that we observed during our study of students' and teachers' behavior and discourse. This model is an attempt to describe the ideological system of the community being studied, and this ideological system may represent the ideology of the international scientific community to which the community under study belongs.

The internalization of a common ideology is especially important in the socialization of scientists, since the scientific community generally lacks national, ethnic, or cultural boundaries (Amsterdamska 1987, 21; also see Stolte-Haiskanen 1989, 415–33, on the need to create and maintain boundaries in the scientific community). Ideally, this community shares a common objective: the search for knowledge using the scientific method, which seeks to transcend political or racial ideologies and other differences. Moreover, the scientific community does not have formal authorities or mechanisms of control over its members. Thus, for example, there is no clear mechanism for gaining admission to the community. Merely having a diploma or academic degrees is not sufficient for an individual to qualify as a scientist; rather, this qualification is gained through the recognition by others of one's scientific product (Hagstrom 1975). This implies, in addition, that the individual must participate in an exchange network with other scientists: she or he must be read and cited, just as she or he must recognize the work of others; this identifies the individual with a community. Thus, the lack of formal controls necessitates the internationalization of an ideology to establish the ideal behavior to which individuals who share it are to aspire, even if in reality they do not behave completely in accord with the ideal.

Science's ideological system includes a certain cosmology, and it defines the role that science has in maintaining this cosmology (Clark 1980, 6–8, 15–25). It contains a system of values and beliefs related to science itself and to the traits that are desirable in scientists. Some ideological conceptions of the relationships among science, the universe, and man can be expressed as follows: man can be familiar with and control science; science and the scientific method are the instruments that allow man to understand nature; science is the sole knowledge system for truly understanding nature; scientists' most important goal is the search for knowledge, and it is from this endeavor that scientists obtain their greatest gratification; knowledge of the universe is morally good and can save humanity; knowledge and the search for truth transcend the individual and constitute an achievement for the entire community. As Bernard and Killworth write: "In our view, scientists make up a distinct subculture in our society. This subculture has its own norms, values, folklore, mythology. 'Knowledge is good for mankind' is the title of our pervasive myth in science. The fact that this is part of scientific mythology does not mean it may not be *true*. A myth is, after all, a commonly believed, commonly (even automatically) expressed rationale for behavior in a group. It is, in fact, the commonality of mythology that gives solidarity to a culture or subculture" (1977, 267).

The ideological system also includes a series of values that make up a model or myth of what the ideal scientist should be like. Myths are a language that give a symbolic meaning to reality. In the case of science, they represent the expression of the ideal of scientific culture, irrespective of its concrete fulfillment. Myths serve not only to guide one's own action but also to educate and train others and to appraise the behavior of the other members of a community. Likewise, they solve, at an ideological level, the contradictions or conflicts of real life (Lévi-Strauss 1968, 198; Lévi-Strauss 1969, 29–40; Malinowski 1971, 44).

The myth of the ideal scientist is not organized into a coherent story, as in the case of primitive societies, not even at the conscious level of most scientists. From our field experience, we emphasize the following characteristics of the ideal scientist:

1. A scientist is a person committed to science, to which he or she devotes his or her life. A scientist must be both disciplined and creative. Discipline is the prerequisite for attaining creativity and, once attained, is the means by which new ideas are processed and communicated until they become part of the scientific discipline or tradition.
2. A scientist must have succeeded, through work and effort, in controlling and managing his or her mind, emotions, and physical energy. Discipline, however, is viewed as a mere instrument at the service of creativity, which leads to the scientist's goal: to produce a new idea.

These two elements—discipline and creativity—are embodied in two key concepts: the scientific method and creative thinking. The "scientific method" is believed to give one the power to become familiar with nature; it is an instrument that the scientist has and uses at will.[1]

---

1. For example, Prigogine (a Nobel laureate in physics) feels that only the scientific method permits humans to establish a dialogue with nature, especially through the experimental

"Creative thinking," in contrast, is conceived of as a gift that one utilizes; although it cannot be foreseen, the scientist seeks to control and exploit it to the utmost. The ideal is that an individual will learn to control his or her creativity.

The foregoing elements of the myth are present in the characteristics that professors try to transmit to their students, as systematized by us in the following model of the ideal scientist:

Control Structure
- Work discipline: efficiency and order; diligence; tolerance for working long hours
- Mental discipline: acquisition of the language, the paradigms, and the method of science
- Emotional controls: tolerance of frustration; control of aggression; independence; patience and perseverance; honesty and trust

Liberating Processes
- Release from the discipline of work: tolerance and enjoyment of disorder; breaking of routines; lack of constancy
- Creativity and release of the imagination
- Release of emotions: 1) motivational stimuli—the desire to do research and the enjoyment of research; 2) self-protecting feelings—the confidence and self-assuredness to undertake activities, defend one's own ideas, and criticize others' ideas.

The ideal model developed here is structured on the basis of opposition to or complementarity with two tendencies that, because of their characteristics, we have called "controlling" and "liberating": the first refers to disciplinary aspects and the second to aspects that stimulate creativity and originality. The internalization of a scientist's social role will imply, first of all, a slow process of acquiring certain work habits, discipline, modes of thought, and emotional controls. Nevertheless, these controlling tendencies would not be sufficient to train scientists since scientists need to produce original work using their own ideas and personal style; these latter qualities allow science to progress and introduce change, and they make it possible for scientific thought to evolve (Lomnitz and

---

method: "The experimental method is central to the dialogue with nature established by modern science. . . . [W]e believe that the experimental dialogue is an irreversible acquisition of human culture" (Prigogine and Stengers 1984, 43–44).

Lomnitz 1977; Collins 1985). Hence, new ideas can be generated at unexpected moments of free association, although invariably they are the result of a prior disciplinary process that allows the scientist to recognize or ignore a question and to integrate new results in the scientific current with the required level of methodological sophistication. The liberating processes will permit the emergence of creative questions or new answers, which in addition must be reprocessed through discipline in formal terms to become part of the scientific tradition. In this manner, the control structure and the liberating processes can continually interact in the scientific language, method, and paradigms, thus molding creative ideas in a formal discipline.

Control Structure

Work Discipline

Valued work habits include organization, a systematic and meticulous focus, a willingness and ability to devote long periods to one's work, and the efficient utilization of time and resources. Perseverance and the ability to work in an independent and self-sufficient manner are also valued.

Mental Discipline

An essential aspect in the training of the ideal scientist is the development of a discipline of thought. All scientists must learn the language—the vocabulary and grammar—of their discipline; in addition, they must learn to think in accordance with predominant scientific paradigms. An important part of the control structure refers to the acquisition of existing scientific knowledge; over time, this knowledge and language become the *foundation* of the structure—that is, the mold that shapes the content of scientific thought through symbols and meanings (language, concepts and rules, theories and techniques) that constitute the elements of scientific reasoning. This mold does not totally determine scientific thinking, although it does shape a researcher's observations, analyses, and interpretations of the segment of reality that she or he is studying, by defining the boundaries thereof and separating her or him from reality as a whole, defining "the logical and grammatical traits of our diverse scientific languages, how we see the world or what we understand to be the facts of the world" (Hanson 1969, 181–83; quoted in Mulkay 1979, 35). In this manner, scientists gradually construct their knowledge of reality (Berger and Luckman 1976; Mulkay 1977; Latour and Woolgar 1978).

The control structure also includes an ability to develop forms of logical reasoning and to formalize ideas. A "natural genius" will not become a scientist unless she or he acquires this ability. Ingenious ideas must also be formulated in terms of an understandable language; in fact, for these ideas to emerge, a language must already exist. Logical reasoning is the ability to set into motion orderly thought processes that can be used to test and analyze ideas clearly and systematically. Logical thinking implies a convergence of thought processes (Guilford 1964) that analyze and synthesize data or ideas that, in the end, will allow the scientist to construct a model or abstract explanation of reality. Logical thinking, however, should be balanced by such liberating processes as skepticism, imagination, and independence from traditional ideas.

The process of thinking logically appears as a pendular movement between two levels: the "concrete-real" level and the "abstract-conceptual" level. It entails inductive, deductive, and analogical thought processes. It includes 1) locating the problem; 2) developing hypotheses from data, with or without prior theoretical models; 3) solving the problem through logical reasoning; and 4) interpreting the data, which leads to modifying the hypothesis, concepts, or ideas. So central is this form of reasoning for scientists that it has been labeled the "scientific method." It is a language model that molds thought throughout the scientific community; it is a type of language or metalanguage through which reality is perceived and explained.

In reality, scientists rarely perform all these processes systematically; they generally skip some steps or perform two or more levels simultaneously. A single observation usually influences the formulation of a problem, and one may arrive at a conclusion without having in mind a clear hypothesis, or one may have the answer to a question before the end of the experiment. In addition, there is a certain division of labor in the scientific community, between experimentalists and theorists. Each of these types of scientist tends to specialize at a certain level of scientific method.

Nevertheless, the final results must always be reported in the manner prescribed by the scientific method, especially among experimental scientists, since only in this way can the quality of a project be acknowledged. This acknowledgment, in turn, gives the researcher recognition and acceptance as a member of the scientific community. No other procedures exist for admitting or excluding an individual from the scientific community.[2] As one of our respondents said: "A scientist is a person who uses the scientific method."

Emotional Controls

Mental discipline must also serve to control the ideal scientist's emotions. Most scientists tend to feel emotionally connected with their study object, and they base their work to a certain extent on intuition, enjoyment, and other affective elements. Such elements are recognized by scientists as a motivating force of science (Cereijido 1991; Prigogine and Stengers 1984; Mitroff 1972; Watson 1968). In the group being studied, these elements were considered to have a transcendental importance; professors tried to detect and stimulate such elements, especially those that are most important to the creative basis of scientific activity. Nevertheless, at a given moment it is necessary to control these elements, to make rational what is mostly emotional, or to defend through the scientific method what is in essence an intuition or hunch. That is to say, scientists, in order to be scientists, need to translate at some point the emotional aspect of their activity into the language of scientific discipline. Scientific thought or the scientific method will, among other things, fulfill the function of controlling emotions, providing them with a rational path to follow. Through logical thinking, then, subjectivity can be incorporated into scientific work. Nevertheless, in formal discourse "avoiding the interference" of subjectivity is stressed: a certain distancing from the study object is considered desirable. Moreover, it is felt that

---

2. See Merton 1973a, Hagstrom 1975, and Collins 1985 on the importance of recognition in the life of a scientist.

scientists must recognize the limitations that arise from their theoretical framework or methodology. In other words, a scientist must be objective, even regarding the degree of objectivity that can be attained. In a broader sense, objectivity must include the ability to formulate a fairer judgment, one that does not exaggerate one's own view, as well as a willingness to acknowledge the viability of one's subjective view when it is found to be valid.

Another important element of control for a scientist is related to frustration. Research generally takes a long time to be completed. It does not consist of a regular production of data and results leading to grand ideas; on the contrary, new ideas take time to be produced and, in the case of experimental science, to be tested and interpreted. Errors cause delays and alter the course of events. Indeed, all scientists go through periods of uncertainty regarding the value of their contributions; these doubts are not cleared up immediately; they frequently take years to be confirmed or discarded. The consequent feelings of frustration, insecurity, and disappointment may cause a serious career crisis. As one student said at the end of his first year in the undergraduate program: "I feel that the undergraduate program is designed for perfect students, [and] . . . that scares me. I'm not sure if I'll be able to [do] this and whether I should be here or not. I'm not sure that I'll be able to become someone who is valuable, rather than a mediocre [scientist]. I would be willing to push myself a lot, if I only knew that I would become good" (I/2nd).

Scientists must develop a high tolerance for frustration in order to be able to carry out their work. They should be able to postpone rewards, both in regard to obtaining final results and in regard to enjoying academic and social recognition. Although this does not mean that they have to forfeit all rewards until the end of their research, it does mean that they should derive pleasure in their daily work in order to maintain an adequate level of motivation, since external stimulus is generally inadequate, intermittent, or slow to arrive.

The search for difficult or at times unattainable objectives provokes anxiety and, occasionally, depression—states of mind that the scientist must learn to tolerate and control, without being inhibited by them. Frustration also increases aggressiveness and anger, which scientists must learn to channel as intellectual, constructive aggressiveness. Another problem is envy, which may arise as a result of seeing in other candidates qualities that one lacks. These feelings must be controlled in order for one to maintain his or her own level or output, as well as for the scientific group to maintain its solidarity, and to ensure the legitimacy of the scientific product. New discoveries, rather than producing jubilation, may induce a feeling of threat or disadvantage in a rival scientist and the consequent appearance of envy and aggression. There are ideologically acceptable mechanisms with which to face and express envy on an individual and rational level. These mechanisms include skepticism, a critical attitude, and competition. Ideally, they must all be expressed in ways that are accepted by the scientific community: intellectual aggressiveness; criticism founded on scientific, rather than personal, bases; and imaginative positions regarding new lines of research.

Scientists are expected to be hardworking and extremely dedicated—to live surrounded by and for their work. For this reason it is important in a scientist's training that a deep commitment to work stand out time and time again—that scientists learn to measure their self-esteem from the appraisal they make of the quality of their endeavor. Ideally, the scientist should learn to appreciate—continually and consistently—his or her own work as well as that of others, and to be aware of his or her own deficiencies and limitations in solving specific problems, so that when another scientist is successful where one has failed, one will not feel resentful or discouraged.

Lastly, one of the fundamental values of scientific ideology is honesty. The ideal scientist refrains from altering or falsifying data to make them agree with her or his own scientific or political ideas; she or he avoids borrowing others' ideas or data; and she or he avoids hiding from the existence of other data that disprove her or his own work. On a deeper level, honesty means not retaining or discarding information out of opportunism, and never omitting critical experiments in testing one's own hypotheses. Likewise, a scientist must trust her or his colleagues' data. Failing to trust others' findings makes scientific progress slower. In reality, there are not always ways to monitor the honesty of the scientific community (Chubin 1989; Kerr 1989; Pinch 1979, 332–33), but the general assumption is that honesty will prevail as a basic value and a social norm respected by the community. Transgressing this norm is decisive: it determines exclusion from the scientific community.

Liberating Processes

The control structure presupposes the existence of impulses that must be channeled. We call "liberating processes" those processes that unleash the flow of forces at an emotional and intellectual level, and at work. These impulses should not be considered to be of a purely instinctive or biological nature but, rather, as energies that have been shaped by a socialization process. Thus, ideas do not emerge in a vacuum, they emerge from existing knowledge; and emotions are felt or expressed in socially accepted ways.

Liberating processes are related to the generation of new knowledge, just as discipline concerns the accumulation and transmission of scientific knowledge. The generation of new knowledge is the cause of change and progress in the scientific tradition, and it must be controlled and utilized by the scientist at the appropriate time. Ideally, a scientist is not a disciplined robot, but a creative person.

Release from the Discipline of Work

Just as it is important for scientists to acquire a work discipline and to be organized, hardworking, efficient, and methodical, they should also be able to break away from this routine and control. The community does not require them to adhere to a monotonous, fixed schedule, as is required in other occupations. The scientists whom we studied devoted more than the necessary amount of time to their work (ten to twelve hours per day), thereby fulfilling the ideal expectation of allowing themselves to be guided by internal controls. It is generally felt that scientists must be able to devote themselves intensely to their work when necessary. Even when a scientist appears to be idle, she or he is expected to be contemplating new ideas, studying new lines of research, or resting from a very intense project.

Some of the scientists whom we studied tended to be undisciplined and disorganized in their work, even when resting, and they generally disliked complying with rules that might constrain them. These peculiarities are widely accepted by the ideology, so long as they are felt to be related to a scientist's idiosyncratic work discipline. But why are they permitted? We believe this may be related to values of individuality. There is the expectation that scientists will have their own life-style, the eccentricity of which is an acceptable and positive trait. In other words, individualism is rewarded. Tolerance of scientists' eagerness to maintain their own personality or style, combined with the expectation that researchers will be capable of freeing themselves from their internal controls, is conditioned to the ideal existence of internal controls. The relaxation of these controls must be temporary, to the point that an excessively long period of relaxation will produce anxiety and the desire to return to work.

Creativity and Release of the Imagination

Creativity has been defined as the ability to produce something new (Barron 1965, 3) or to recombine, associate, or synthesize already-existing elements in unexpected ways through diverse thought processes (Guilford 1964, cited in Getzels and Jackson 1973, 771). It has also been described as the ability to discover new or hidden facts, by deducing them from missing information, or to detect an epistemological obstacle (Bachelard 1979, 765).

There exist different styles and levels of creativity among scientists. Gough (1961) points out that there is a cognitive preference for the type of complexity (wealth of content, ambiguity, asymmetry, and dynamism) that distinguishes the creative personality in general and the creative scientist in particular. Cognitive flexibility is a characteristic that has been consistently related to creativity, since creative minds move easily from a secondary, intellectual, complex process to a primary, unconscious process in which fantasy, imagination, and ingenuity prevail. This frequent shift between primary and secondary processes is possible because of an ego force and good control over the processes. A creative individual's cognitive characteristics also indicate the existence of a perceptual opening toward internal and external stimuli, curiosity, intuition, and imagination. The scientist tinkers, looking for questions more than for answers. Cognitive elements interact with emotional elements, including personality traits, motivational needs, and emotions, as part of the creative process (see Dellas and Gaier 1970; Koestler 1975; Kuhn 1977).

Our respondents generally considered creativity to be a mysterious phenomenon inherent to the individual and difficult, if not impossible, to teach or bring out. Stimulating and developing creativity is one of the essential interests of the professors.

With respect to ideology, creativity and the search for knowledge symbolize the ideals of science. Carrying out creative work stimulates the scientist to continue with his or her efforts, despite the rigors of a research career. In the creative process, the imagination is freed and a certain restraint on the unconscious is removed; the individual sheds certain intellectual controls, thereby making it possible to think in a different or new manner.

Scientists usually describe their creative experiences as moments occurring at a low level of consciousness. The nature of these moments is more intuitive or unconscious than rational (Cannon 1965; Szent-Gyorgy 1962). The sparks that free one's intuition and imagination are not created in a vacuum, though: on a preconscious level they arise from existing knowledge and must be converted to this knowledge, as the new ideas are registered within the framework of given disciplinary controls. One professor told us:

> Being too methodical is detrimental—always wanting perfection, checking the literature to the last detail, and having perfect control.... There are times when you have to get away from this, to stop being methodical ... to stop going to the library to read everything that has been done in the past fifty years; because if you don't, you come to a point at which you can't do anything new. You do only what everyone has done or taught before.... When you begin to do things without a particular method and you arrive at a concept that is somewhat new—that is the time to go back and do it right. The idea is to know when to get out; *that* is the important point in the maturation of a scientist. (I/3rd)

Release of Emotions

It is felt that certain emotions need to be controlled or channeled; for if they are not, they will obstruct the process of scientific production. In this category are envy, frustration, aggression, anguish, and depression. Other emotions, however, need to be freed so that they will stimulate and invigorate the scientific process. In the next section we will analyze the values relative to emotional currents that should be allowed to flow and that will help scientists tolerate the long periods of monotonous discipline they must go through, all the while allowing them to conserve their energy in order to produce bursts of creative work. We have identified two sets of what are assumed to be driving emotional currents.

Motivational Stimuli

Human beings are motivated by desire, which can materialize and be satisfied in different ways. Nevertheless, one never fulfills or satiates this desire (Lacan 1976). It is assumed that scientists have an underlying desire to learn, to see and to know what is hidden in nature. The search for knowledge is in itself inexhaustible and, therefore, can never be totally satisfied. This translates into the scientist maintaining a level of motivation that becomes a constant search for more knowledge. It is this constant search for knowledge—rather than the production of knowledge—that is assumed to be satisfying and stimulating in itself.

So long as the scientist produces new knowledge, new questions will arise; thus motivation will be maintained. In the life of the scientist there will be periods of enthusiasm when one feels that a new research vein is being opened, when one's hypotheses turn out to be correct, and when one receives recognition. Nevertheless, scientists are expected to be able to control their enthusiasm and not to let it interfere in the development of their research. One professor said: "When [the members of my research group] are depressed, I raise their morale, and when they are too enthused, I bring them down to reality" (I/4th).

Group work, cooperation, and even competition also constitute motivational stimuli for the ideal scientist. As one student said: "To work in a group [with all of us] together is really neat; you feel that everyone puts in their grain of salt to solve a science problem; . . . it's very exciting. You worry that other groups will win the game, after you've spent some time thinking or working on a problem. . . . It's like a race; you have to rush" (III/3rd).

Ideally, research activity and the enjoyment it produces constitute emotional stimuli. The previous section, on controls, may seem to have depicted researchers as insensitive, tedious persons who receive little satisfaction and who learn to postpone gratification. But this image must be complemented with a portrayal of the true enjoyment the ideal scientist finds in her or his work. To give free rein to one's curiosity, to search for questions and answers, to feel that one is brilliant and intellectually productive, to obtain results and see them published and, in the case of experimental researchers, to observe and manipulate reality in the laboratory, these are considered sources of pleasure, as are the personal relations a scientist may establish with colleagues, students, and teachers. (See autobiographies by scientists: e.g., Cannon 1965; Snow 1965; Szent-Gyorgy 1962; Cereijido 1991.)

Self-Protecting Feelings

Self-protecting emotional mechanisms include self-assurance and confidence in oneself, in one's ability, and in the potential of science as an effective method for acquiring knowledge. Self-confidence will sustain the ideal scientist in times of frustration, doubt, and uncertainty. What is more, scientists must be able to trust their own intuition and believe in their ability to see and understand things on their own, as well as to find their own ideas; in short, they must trust their intellectual abilities. At the very least, scientists must have sufficient self-confidence to submit their results for publication and to speak in public in front of their colleagues. This confidence is related to the values of independence and creativity. Self-confidence will allow one to defend her or his own ideas, to criticize those of others, and to persevere in the face of failure.

In summary, scientists must face what are ideally two contradictory tendencies: the impulse toward creativity, fantasy, subjectivity, and freedom of thought and, at the same time, the development of an internal control structure that shapes and limits this impulse within a coherent, logical, and manageable system of thought. In reality, these tendencies interact in different ways and to different degrees, in concert with the scientist's personality. Some scientists tend to work by following their impulses, intuition, and fantasy; others tend to be more disciplined. Each scientist chooses to develop her or his own ways of managing these two tendencies, since both must be used at different moments over the course of a research project. Kuhn (1977) talks about the "essential tension" necessary for a scientist, and Merton (1976), using a different framework, mentions "ambivalence"; both authors refer to the existence of these two opposing currents that scientists must handle. Kuhn (1977, 226) defines "divergent thought" as the freedom to select different approaches, to reject old solutions, and to venture in new directions; he understands "convergent thought" as that thought that has been integrated by and transmitted through the scientific tradition. Since these two types of thought inevitably conflict with one another, one must be able to tolerate a certain amount of tension, which at times can become excessive but which "is one of the primordial requirements for the best type of scientific research" (Kuhn 1977, 226).

The scientific life is a search for internal order in nature, and scientists tend to construct a symbolic or model system that will explain nature. Paradoxically, this search for a theoretical formulation or for an explicative model of the universe leads scientists to create a type of disorder in their own formulations, as they recognize inconsistencies, ambiguities, or flaws in their explicative models. Both of these—order and disorder—are necessary for the production of knowledge: order serves to store knowledge that has been acquired, while disorder allows new questions and ideas to emerge (on this point see Latour and Woolgar 1978, 235–61; Knorr-Cetina 1982; Collins 1985).

Ideally, the would-be researcher will develop a control structure or a discipline and, at the same time, will learn to release her or his creative tendencies. In this process, she or he will also learn to face the internal contradictions these two tendencies produce. On the ideological level, this dynamic process is reflected in the specific behavioral reality of each individual, and it determines one's style as a scientist, in accordance with one's personal characteristics and those of the teachers who most influenced one. The process of assimilating the normative and ideological aspects of the role of the scientist, which is carried out via interaction with teachers and individual integration, determines whether the student will acquire an identity as a scientist.

The scientific community seeks to reproduce itself by training individuals who adhere to its ideology; this underlines the originality and importance of the scientist as an agent of change. Hence, both continuity and change are sought, and for this reason scientific ideology is complex and, at times, contradictory.

It is important to stress, however, that the model described here is an ideal and is transmitted as such, which does not imply that it translates literally into individual behavior. Scientists gradually develop their own model, an ideal of the ego, which serves as a guide and an evaluation system. This ideal is rarely acted out in its totality, but it must be acted out partially. Moreover, the ideal model integrates polar, contradictory traits, thus allowing the individual to integrate some qualities in accordance with his or her personal characteristics.

Scientific Ideology in the Literature of Sociology

R. K. Merton has studied the emergence of science in the seventeenth century and the influence of religion, especially the Puritan ethic, on its development. As for methodology, Merton based his work on numerous biographies of scientists, analyzing their successes and failures and the type of recognition given them by the scientific community (Merton 1938, 1973a, 1973b, 1976). He stressed the role played by the norms and values, or ethos, of science in the formation of scientists:

> The ethos of science is that affectively toned complex of values and norms which is held to be binding on the man of science. The norms are expressed in the form of prescriptions, proscriptions, preferences, and permissions. They are legitimized in terms of institutional values. These imperatives, transmitted by precept and example and reinforced by sanctions, are in varying degrees internalized by the scientist, thus fashioning his scientific conscience or, if one prefers the latter-day phrase, his superego. Although the ethos of science has not been codified, it can be inferred from the moral consensus of scientists as expressed in use and wont, in countless writings on the scientific spirit and in moral indignation directed toward contraventions of the ethos. (Merton 1973a, 268–69)

Merton points out four basic "moral imperatives" that he feels are the foundation for social relations among scientists and for their professional identity: "universalism," "communism," "disinterest," and "organized skepticism." He distinguishes two types of norms: 1) technical or methodological norms and 2) moral or ethical norms. Together these work to increase science's primordial objectivity—the production of "certified knowledge." In Merton's words: "The mores of science possess a methodologic rationale but they are binding, not only because they are procedurally efficient, but because they are believed right and good. They are moral as well as technical prescriptions" (1973a, 270). In this manner, ideology (embodied in moral norms) is able to permeate the application of methodological norms in the practice of science. Sometimes compliance with these rules will occur freely; on other occasions it will not, as prohibitions will have been internalized. Or, in extreme cases, it will occur through the disapproval of other scientists; and, in the most extreme cases, sanctions will be applied. Barber (1952) later added two norms to the scientist's ethic: individualism and rationality. Storer (1966) added those of objectivity and generalizability.

Merton's ideas gave rise to a current of studies that predominated in the sociology of science during the 1950s and 1960s. Starting in the 1970s, several criticisms of his proposals appeared, leading to the abandonment of studies related to the ethos of science—and, therefore, to what is described by scientific ideology—and its role in the formation of scientists (see Barnes and Dolby 1970; Mulkay 1980; Mitroff 1974).

Criticism of Merton is divided into two types: 1) those criticisms referring to scientists' compliance with behavioral norms; and 2) criticisms of his functionalist approach, according to which compliance with norms is what permits scientific knowledge to move forward. The first research work that criticized the model proposed by Merton was Mitroff's study (1974), which described the attitudes, beliefs, and practices of scientists involved in the Apollo mission during the five years it lasted. Mitroff did not find the scientist's emotional neutrality or stereotyped disinterest; rather, he concluded that science advances because of passion in scientific work (Mitroff 1974, 24). Science, according to Mitroff, is a human, individual game that requires a personal commitment and that cannot cease to be, to some extent, subjective: "The game has powerful subjective elements . . . [and] strong irrational components. . . . This does not mean that it is totally subjective, irrational, and relativist" (Mitroff 1974, 268). Moreover, he feels, not only are Merton's norms not complied with, but the counternorms (such as subjectivity or an interest in advancing one's own ideas) are necessary for the progress of science: "The scientist is to be 'committed' to [and at times even 'prejudiced' in favor of] his favorite theory, his hypotheses or positions if he wants the scientific community to pay attention to him. In sum, if the scientist does not *advocate* his theories . . . he may be . . . ignored" (Mitroff 1974, 614).

Mahoney (1979, 349–75), basing his work on biographies of scientists and on archival material, attempted to ascertain whether the "ideals" of science coincide with the scientists' actual behavior. He analyzes five of the most widely accepted values: objectivity, rationality, open-mindedness, superior intelligence, and integrity. He shows, with examples of scientists' conduct, that actual behavior does not always comply with the ideals. Mulkay (1979, 224–57) also criticizes Merton, on the grounds that what scientists say is different from what they do, and that there is no evidence that scientists believe in the values ascribed to them.

Barnes and Dolby (1970, 3–25) criticize the norms proposed by Merton, which they consider outdated and functionalist. They argue that these norms represent an ideal for scientists, a myth based on seventeenth-century amateur practice, but that they do not conform to the present-day situation of the professional scientist, who is organized into groups and who competes for funding and recognition. A scientist gives loyalty (not skepticism) to her or his own paradigm; she or he communicates with members of her or his own group (which is not universal); and she or he is not always rational.

Merton (1973a, 259) responded by arguing that his characterization of the normative structure of science refers to an ideal level, not to its application or practice in scientists' daily conduct. He says, for example, that values such as organized skepticism, disinterest, and impersonality do not necessarily imply that the scientist is a cold person; on the contrary, the scientist invests considerable emotion in his or her way of life. Merton also observes that science is characterized by potentially incompatible normative demands (norms and counternorms) and that the scientist who has internalized these values goes through periods of tension and conflict. He considers this "ambivalence" useful and necessary (Merton 1976, 32–64).

The second criticism is influenced by Thomas Kuhn's theoretical development. Kuhn postulates that the scientific paradigm is a source of social control and that the acquisition of social norms is measured by the conceptual structure of science (1962, 46). Merton's most important critic, from this perspective, is Mulkay (1969, 1976, 1977, 1979), who feels that Merton's approach is the result of a vision of science as descriptive of true knowledge, a vision that assumes that once the sources of distortion are eliminated, it is possible through systematic observation to recognize empirical regularities in nature. For Mulkay the norms proposed by Merton attempt to reduce these possible distortions in scientists' behavior and thus promote the advancement of knowledge. Mulkay proposes a new scientific vision, which he does not conceive of as being aimed at a representation or description of reality but, rather, as one in which reality is constructed intellectually and socially through a creative process of the production of knowledge, in which new knowledge with new meanings is created (1979, 64–65).

Mulkay questions the effectiveness of the "ethos of science" as a valid way of distinguishing science from other institutions, and he suggests that a differentiation be based on a theory of the development of specialized knowledge. He maintains that "theoretical and methodological norms are more important in the structure of the scientific community than are Mertonian social norms" (Mulkay 1969, 36). For Mulkay, the scientific community's identity is cognitive or intellectual; he feels that shared theoretical knowledge and the methods used to generate and guarantee knowledge are important even in the scientists' social behavior: "Thus it is argued that scientific propositions portraying the physical world also constitute standards defining how researchers are expected to perceive the world and how they ought to undertake their research" (Mulkay 1977, 146).

In the 1970s, then, the anti-Mertonians attempted to refine the sociological analysis of scientific knowledge, for they felt that norms did not completely explain scientific production. In the 1980s, the sociology of science began to make a turn, with the internalist focus prevailing. Now microsociological studies of scientific practice were privileged: there was a tendency to give priority to the question of how scientists go about talking and doing science, of why they act as they act; the tendency was to adopt what can be called a "constructivist perspective." The constructivist perspective is characterized by a concern for the process by which outcomes are brought about through the interactions of individuals. This perspective assumes that outcomes are the result of the participants' interactive and interpretative work, that social processes are constitutive of the production and acceptance of knowledge claims (Knorr-Cetina and Mulkay 1983, 89; Cozzens and Gieryn 1990; Latour and Woolgar 1978; Lynch 1985).

With this focus, ethnographic studies concerning the production of knowledge investigate the very place where scientific activity occurs—the laboratory—considering scientific fact as the product of a social construct starting from the interaction of scientists. This focus goes so far as to consider scientific facts as strictly constructed and the acceptance of those facts as the outcome of entrepreneurial ability and power on the part of the scientist and her or his network.

By the 1990s, there was a reaction to constructivism. A balance was sought between those systems that are internal and those that are external to science: an analysis of the development of knowledge per se, in which the continuity of science consists and which is what distinguishes the scientific community from other social groups, and scientific activity from other activities; an explanation of the " 'why' of scientific tasks, as well as the 'how' " (Hagendijk 1990). Even Latour (1991), one of the initiators of the constructivist current in science, acknowledges that in his desire to rebel against the Mertonian school he came to view scientific activity rigidly, concentrating on the social aspect of the activity and neglecting its nature (see Bunge 1991 for a critique of the constructivist school; also Vessuri 1991 and Doran 1989).

The controversy has not ended. Hackett (1990) proposes a framework for thinking about value change in academic science: culture is to be viewed as a set of competing variables, and academic science is to be considered an organized activity shaped by organizational forces, particularly by the quest for resources and legitimacy. Hackett feels that norms and counternorms alike are part of a cultural system that influences science by means of the different social contexts within which science operates. In the case described by Mitroff, for example, the competitive and politically charged atmosphere of space science research in the early 1970s brought the labeled counternorms to the fore (1974). Academic science, on the other hand, is carried out under different organizational conditions, which in turn are affected by external forces such as state or societal pressure, which produces ambivalence and tension. That is to say, conflicts may result from the different social or institutional conditions in which science occurs, and not from real differences having to do with the existence or nonexistence of norms and counternorms. Nevertheless, an ideology that affects scientists, giving them identity and cohesion, prevails in the scientific community as in every social group.

Although there has been much discussion of the normative system and its effectiveness or ineffectiveness in the functioning of the scientific community, social scientists, with few exceptions, have not been eager to conduct research based on empirical studies of the problem of training new scientists (Bucher and Stelling 1977; Reinhartz 1979). Although some studies have shown the importance, for example, of the relationship between students and teachers in the training of scientists (Hagstrom 1975; Zuckerman 1977), no studies have 1) described the socialization of scientists ethnographically; 2) analyzed the manner in which ideology, theory, and technical knowledge are transmitted and internalized; or 3) described how scientists work and what sort of social relationships exist within the scientific community. This study is an attempt to bridge that gap. In our study, although we have used a methodology similar to that used by the constructivists (in the sense that we did detailed work with a small group of scientists), we did not do so in order to describe how scientific facts are produced, but rather to understand the training of scientists through the transmission and internalization of an ideology, with its worldview, norms, and values, analyzing in detail what that ideology is as well as the internal conflicts that arise.

We have observed that ideology, the set of beliefs and values, was a *real* tool of socialization. Throughout the process of training students, the transmission and internalization of ideology was stressed as much as the acquisition of knowledge and methodologies; ideology was an element of community cohesion as well as a guide for action, for scientific work. Still, we see this ideology as a set of values that works at an ideal level, not as a set of normative rules for action. The ideological system contains contradictions and ambiguities that allow the scientist to act in distinct (and, at times, contradictory) ways without feeling that the ideal is being violated. We treat ideology as a system made up of multiple symbolic elements, which can be combined in different ways and interpreted with subtle nuances.

The different values represent an inventory from which individuals select items according to the type of work they are doing and according to where they are in the process of scientific production. Thus, at certain times the scientist must be passionate and subjective in defending his or her ideas, but also objective in analyzing the ideas of others and in presenting his or her own work, if that work is to be accepted by the community. In addition, the inventory of values is transmitted, and students (depending on their experiences) develop interpretative criteria allowing them to act—just as a grammar, through a limited number of linguistic rules, allows each individual to make new combinations and give a unique linguistic performance, all while abiding by the rules. In the following chapters, we will see the role played by the ideology of science in the socialization process and in the acquisition of a scientific identity.

5

The Socialization Process

To acquire an identity as a scientist it is not sufficient to learn a repertoire of scientific knowledge and techniques; it is also necessary to possess a set of values and behaviors shared by the scientific community. Thus, scientists learn to see and evaluate themselves and the world in a distinctive and characteristic manner; in order to do so, they internalize an ideology.

The development of this identity is the result of a long process of interaction between students and researchers, during which students act out the different aspects of a researcher's role. In the end, this leads them to assimilate their self-image as researchers and to be accepted as members of the scientific community through the mutual respect provided by sharing certain beliefs (see Freud 1921). The scientific socialization that we observed in the undergraduate program in biomedical research took place via three complementary processes: 1) the relationship with teacher/advisers; 2) interaction with the group; and 3) role-playing.

The foundation for training researchers in this undergraduate program was the relationship between students and teacher/advisers. Each teacher acted as a role model and guide. Students learned by imitating their teacher/advisers and by being guided by them. They gradually came to identify with the different ideal traits that would govern their conduct. Ideally, each teacher/adviser gave students personal attention, attempting to stimulate the development of their potential, especially originality and creativity (see Collins 1985, 56). Hence, the relationships between teacher/advisers and students were not anonymous or impersonal. Each relationship was unique: "There is a particular form of sociality, for science like other practices is taught and learned. . . . The acquisition of observational techniques, the internalization of a given language, the feel for the right question, the sense of what to do next—all that belongs to the repertoire the novice has to acquire, so that he or she, a scientific learner, becomes in part, a different person. . . . [Thus one begins by] learning from masters, . . . entering into a new intellectual and technical world" (Grene 1984, 14–15).

For the students, the teacher/advisers were the first models of identification. Later, the role models were increasingly abstract: distinguished scientific figures, authors of relevant works in the field with whom the students would become familiar through their reading assignments. In the last stage, students entered a work group, where they remained for a long period, at the end of which they did their undergraduate thesis. At that point they formed a closer relationship with their teacher/adviser, and through this relationship they learned to work by assimilating the values of their teacher/adviser.

The integration of the student into a work group and/or into the scientific community was gradual. Initially, it occurred only in teachers' formal discourse. As the student overcame obstacles and began to be recognized as an individual who shared a common ideology, language, and work method, she or he was gradually identified as a member of a group, which would in turn give the student a feeling of belonging.

Upon entering the undergraduate program, students confronted the image of a scientific community as an exclusive, closed group to which it is difficult to belong. Professors took pains to stress that belonging to this community implies a personal commitment in addition to mutual recognition. Later, it became clear that entering the scientific community is a slow process in which a student attains partial recognition from teacher/advisers and makes gradual commitments. Likewise, students developed their own system of internal control, which motivated them to behave according to expected norms and to attain satisfaction from their work (De Vos 1979, 214). The discourse of teacher/advisers was charged with value judgments relating to the behavior that was expected and relating to what constitutes "good scientific work."

Initially, undergraduate-degree candidates formed a cohort—just like the students in any of UNAM's undergraduate programs. However, the fact that these students belonged to an experimental, selective program distinguished them as a small and privileged group, enjoying access to the Institute of Biomedical Research and to a group of highly qualified researchers. The generation, the undergraduate program, and the Institute became the students' reference groups. When the students joined a specific laboratory, the cohort that made up the laboratory became the reference group and the student ties with previous classmates weakened. In later stages, the students joined a network of colleagues sharing a common interest in specific scientific problems, both within and outside Mexico. Such networks, or "invisible colleges," (De Solla Price 1979; Crane 1972) have a well-known place within the scientific community. Linkage with these networks first took place through the agency of teacher/advisers and through reading assignments, participation in scientific congresses, and contributions to publications.

In the undergraduate program, the active resolution of scientific problems using the experimental method was the most important focus of teaching (see Chapter 3, above). Students were required to adopt the role of experimental researchers, although in fact this role-playing was fictitious. As students learned to do the work of a scientific researcher, they acquired scientific habits, knowledge of the discipline, relevant methodologies, and the norms and values that make up the ideal model for an investigator. Specifically, students were encouraged to seek knowledge using the experimental method: by raising questions or problems, designing experimental strategies to solve them, carrying out experiments, analyzing the results, and raising new questions that would allow them to move forward in their study of a piece of nature (see Chapter 3).

In reality, the three socialization mechanisms are intimately connected and cannot always be separated at each moment of the process. Thus, what has been called the "mystique" of scientific activity is transmitted simultaneously through the acting out of duties, through teacher/adviser relationships, and through group interaction. For example, during the course of experimental work, the professor conveys the formal or curricular elements of the course she or he teaches (see Appendix 2);

however, the professor also transmits a series of ideological elements—criteria regarding the validity of theories and their impermanence; the need to corroborate, perfect, and surpass theories; and the need to have or acquire one's own ideas, to know how to defend them, how to retract them in time, and how to criticize them.

The student's interaction with nature, in the controlled situation of the laboratory, allows the teacher/adviser to objectify the normative messages by means of personal, concrete specifics. Finally, interaction with teacher/advisers, classmates, and other researchers in a context of work, discussion, or rest allows the scientific aspirant to present her or his own ideas by classifying them in terms of new categories ("originality," "self-criticism," "independence of criterion"). In addition, the student learns the language of the discipline, the proper way to attack or defend an idea, and the rules for social interaction. Below, we describe the process of socialization as it developed in UNAM's undergraduate program in basic biomedical research.

The Initial Contact

The process of socializing future scientists began with recruitment. The purpose of this stage was not only to select candidates for the program but to familiarize them with the scientific culture and its ideology and, more specifically, with the model of the ideal scientist; this constituted the basis of their identification with the scientific figure.

The student's first formal contact with the philosophy of the degree program was provided by the brochure describing the program; this brochure was distributed to persons interested in the degree. Something of the formal discourse of scientific ideology can be seen in the following passages taken from the brochure:

> The purpose of this degree program is to train scientific researchers who have aptitudes for analysis, synthesis, self-criticism, and self-training in the field of biomedicine. . . . From the time they enter, students are considered researchers and conduct research on increasingly complex problems. Information-giving classes and survey programs are kept to a minimum. Research is

conducted under the supervision and guidance of more than one professor for each problem. The professors are career researchers; that is, individuals whose lives are [devoted to] research, most of whom belong to the faculty of the Institute of Biomedical Research. . . . In the degree program, study, discussion and, especially, experimentation activities require . . . the student's full-time dedication. (IIBM n.d., 3, 8)

Gaining admission to the degree program is not easy, as it requires overcoming a series of obstacles that gives the degree an image of being something which is difficult and special to attain. In addition to meeting the formal requirements (prior studies, good grades, passing UNAM's general admissions examination and the examination developed by the Institute), candidates for admission to the undergraduate program must "demonstrate a vocation for the field during a period of exploration and training in the Institute of Biomedical Research; possess scientific knowledge allowing them to perform the field's particular activities" (IIBM n.d., 5). This message conveys the importance of demonstrating a vocation for a scientific career: merely meeting the formal requirements was not sufficient.

The brochure also conveyed the idea of a select group—that of the Institute's researchers—as well as that of a small close-knit group, almost like a family, in which anyone aspiring to join had to be carefully evaluated. Individual interviews as well as the structure of the introductory course, with its highly selective nature, and that of the undergraduate program in general reinforced the idea that exceptional students were being sought, and that those who were admitted would be joining a select group; admission into such a group was not a mere formality. Finally, in the talks they gave concerning a career in research, researchers made note of the sacrifices that would be required in order for a candidate to become a scientist; they also pointed out that the goal of the scientist is not to earn money or acquire power, but to engage in the search for knowledge.

The researchers presented themselves to the students as a group of people who identified with a role, who carried out specific work (research), who had their own demands, who set themselves apart from other individuals in society, and who belonged to a community. The special nature of this role was implicit in the message: researchers have a unique life-style; they must be devoted exclusively to their profession;

and they must be disciplined and give their scientific career priority over other interests. Thus:

> I told the [students] in the introductory course that at the end we would evaluate them, but that it was more important that they evaluate themselves and see if they wanted to stay here.... Nor do I want some students to cling to me as if [they were attached] to an umbilical cord, unless it is their conscious decision to follow a [certain] theoretical current. I want individuals to be themselves.... Individuals cannot shield themselves behind anonymity. This is a place for one to be continually exposed to criticism, which is the hardest thing to put up with, and I tell them to be aware of this. At the end of the degree program it will no longer be self-criticism alone, and this will apply for your whole life. (I/1st)

In the next section we will analyze the evolution that took place with respect to the training and socialization of students, based on the three channels of socialization. First of all we will describe what occurred during the first two years; then we will discuss the third and fourth years of the program for both cohorts.

The Relationship with the Teacher/Advisers

Students' relationship with their teacher/advisers is the key to the formation of a scientific identity (Hagstrom 1975; Zuckerman 1977; Grene 1984). This is a dynamic interaction in which both parties continually define themselves. Students shape their own image and structure their identity as future scientists based on the signals they receive through interaction with their teachers (Laing 1969; Watzlawick et al. 1971). Likewise, the teacher/advisers, through interaction with the students, shape their own identity as teachers and fulfill their desire to go beyond themselves; all of this contributes toward producing an intense emotional relationship such as the one that arises in primary socialization processes.

Beginning in the first semester, teachers tried to establish personal relationships with students as equals. Teachers would behave informally, with no expectations of establishing differences. They would ask students to address them using the "familiar you" form (*tú*), and they would show an open and receptive attitude and encourage students to challenge and question their ideas; in seminars they would invite students to consult them whenever they wished. The purpose was to convey the message that the community was an open one of equals in which some members (the students) were merely younger and inexperienced.

Parallel to this message of presumed equality, an opposite message—regarding how difficult it is to become a scientist—was also being conveyed: only when students proved they were equal would they be treated as such. In other words, the community was presented as a warm, open, and fraternal group, yet one that was inaccessible to students until they overcame the testing process and showed that they deserved to be admitted.

A relationship with teacher/advisers does not come about automatically. It is a relationship that is formed slowly and in which the student must gain the teacher's attention and interest, even though the teacher appears to be offering this from the beginning. During the first year, the students perceived this contradiction and did not believe the teachers' egalitarian message, not only because of the clear difference in knowledge between teachers and students, but because they did not understand how they were expected to behave. Teachers would try to force the students to participate and not to fear thinking on their own, to defend their ideas aggressively and to criticize the ideas of others. This was interpreted by the first-cohort students as a message of inequality: "Although the teachers try not to make us feel they are better than us, their manner of interrupting in seminars is very sharp; . . . they have an attitude that 'no one knows anything' and 'I'm the one in charge.' . . . The relationship continues to be one of the big teacher versus the small student" (I/2nd). And a first-year teacher commented: "We still have to act [vis-à-vis the students] like teachers and ask them if they have done anything, and [if not] why not" (I/2nd).

At the end of the second semester, the first-year class began to catch on: "I feel that the teachers tried to teach me to see them not as ogres but as persons; . . . [W]hat I liked about the semester was the atmosphere—a big family of teachers and students" (I/2nd). What students best understood during the first year was the need to be assertive and critical in discussions. An aggressive manner, arrogance, and a lack of submissiveness were considered by students to be important attitudes. Generally it was the most insecure, least assertive students who had more trouble responding to this challenge, and they received greater criticism from teachers until they reacted as desired. At times, however, they would also overreact, such as when one student told his teacher: "I don't believe what you say about the second law of thermodynamics, you haven't convinced me" (I/1st).

The third cohort seemed to assimilate the expected relationship more easily, probably because it was a more confident and assertive group of students and because it was introduced to research work from the beginning: "I like the way they treat you, not as an ignorant student but as an equal who knows a little less. . . . Each researcher is different, but here you can say whatever comes to mind; they don't put you down. . . . I feel great; they make you feel like one of them" (III/1st).

During the second year, teachers conveyed the idea of equality by requiring students to begin assuming the role of laboratory researcher. Having already worked on a research problem, students began to enter the researchers' system of relationships; they felt they were treated like researchers. As the year progressed, students began more and more to assume the researcher role: they would formulate problems and experiments and then discuss them, following the behavioral guidelines set forth by the researchers. One teacher, in reference to a seminar, said: "It was an interesting and fair experience, because I don't know about math and they don't know about rules. We taught each other" (III/3rd).

In addition, teachers acknowledged openly that they were influenced by their students' comments: "I've had to study things that I wasn't clear on or that I had forgotten, in order to refute their arguments without interrupting them in an authoritarian way" (I/3rd). Another teacher said: "I came to class the first day with a problem all ready. . . . I let the students discuss it, and I finally realized that there was no reason to focus them on an idea of mine. . . . So I let them discuss and find the problem themselves" (III/3rd). One student said: "We had one seminar in which each participant brought up whatever came to mind and we would discuss [it]; in the discussion it turned out that [we had] more interesting problems than those brought up by the teacher. He let us keep on reading to see what else we would come up with" (I/3rd).

At this stage, the message of equal treatment transmitted confidence to the students as well as a recognition of their potential to become researchers. But the students also had to learn the formalities: it was not a simple matter of being aggressive, or overlooking their teachers' experience or authority.

The Demands

The message of how difficult it is to become a scientist was conveyed, in addition, through the stringent demands that professors explicitly and implicitly placed on students. For example, the students were requried to work long days and had many work obligations, which implied that they make a commitment to dedicate themselves exclusively to their studies: scientists must dedicate all their time to their work; they must be devoted to research. One professor said: "Students have to learn the need [that they work here] full-time—for long, continuous periods—that they be devoted to school activities. . . . The problem that arises in the beginning is that they must internalize this; . . . problems of adaptation to the schedule and to the intensity of the work emerge. They must study all the time, not only during exam [periods]. They must learn to be confined to a single location" (I/1st). Professors discouraged students from making commitments away from the Institute, such as pursuing hobbies or becoming involved in other extracurricular activities.

Teachers demanded from the beginning that students behave as "small researchers"; that is to say, that they show they are capable of doing research and that they have the qualities, if only in an incipient way, of the scientist. Hence, students were required to show themselves to be hardworking, efficient, creative, and critical—and they were not always successful. One professor told us: "One mistake of mine was to assume that the students had the attitude of researchers" (I/2nd).

Requirements were transmitted through the demanding syllabi and through expressions of dissatisfaction and criticism, so that students would produce more in their various activities. Fatigue, anguish, and frustration were the students' typical feelings during the first and even the second semester. They felt the pressure and dissatisfaction of their professors and were anguished and insecure at not being able to understand and fulfill their professors' demands: "I felt very pressured by [the professor], not because he said anything to me, but because I feel the professor expected something from me . . . and that I wasn't responding" (I/2nd). Such pressures induced the students to work hard and to begin discarding activities not related to research: they would even remain at the Institute more than the required number of hours. Still, their anguish did not diminish: "I learned a lot, even though I did not appear to produce what they wanted me to produce" (I/2nd). Stringent demands were something that the students were constantly aware of, and this was noticed by the professors: "That's what I haven't liked; . . . I feel pressured to be aware of everything the professors say or do. . . . They're always testing us" (III/1st).

Toward the end of the second semester the students no longer resented being pressured, and they began to attain satisfaction in their work: "[The second semester] was better than the previous one: there were better subjects; there was more freedom to do things; we didn't feel pressured about time [there was no longer a fixed schedule], although the professors now pressure us as much or more than before" (I/2nd). We feel that the stringent demands placed on students that first year introduced them to an intense work routine, which indirectly conveyed to them the image of the ideal researcher. Students had to work at a fast pace, tolerate a high level of anguish and frustration, and learn to live in a new environment.

In the second year the demands were equally stringent, although they were less obvious. Each professor and her or his group would stimulate the students to work hard and to obtain results from the experiments, and they required the students to work on problems at a high academic level (far beyond the problems worked on by students in traditional programs) using a complex methodology. One result of this socialization process began to be seen toward the middle of the second year, in which the students devoted all their time to their experiments and remained at the Institute all day, and even at night, when necessary. The lack of freedom and the intense pace of the work came to be considered normal. One student said: "I don't see people; I'm stuck here all day; I work weekends [when I have to do an experiment]; I don't go out partying any more; . . . [D]uring the week, too, I stay until late. Once we went home at 2:00 A.M. Every day I stay until 9:00 or 10:00 at night. Sometimes I would go home at 9:00 P.M., and at 7:00 in the morning I was back working because if I didn't, we wouldn't have time to do the experiments. You get home half-dead and go to sleep" (III/3rd).

Independent Work

The students' relationship with their teacher/advisers brought together two apparently contradictory tendencies: the teacher/advisers' role as the students' guide and the encouragement they gave the students to work independently. As explained in the preceding chapter, an important aspect of the process of socializing scientists is that they learn to be self-sufficient in locating information, generating ideas, and designing experiments. The teacher/advisers went to great lengths to allow students to work on their own and to try to solve on their own most of the problems they were assigned. This does not imply that the professors actually left the students alone, even though the students did perceive it that way.

For the students, working on their own represented a change vis-à-vis their prior education. One professor said: "A problem we have with the students is that they have not taken a responsible attitude regarding the undergraduate program: we still have to function as teachers, ask them if they have done something, why they haven't . . . and be on top of them" (I/1st). Or, as another professor said: "The students still have a traditional attitude regarding the teachers; they feel frustrated because we don't want to direct them as [is done] in [traditional] schools" (I/2nd).

The professors expected students to manage on their own and to learn to make decisions; in other words, not to depend on their teachers. One professor said: "I try to teach them to look for the information themselves . . . to think and decide on their own, and to ask many good questions: [questions that are] clear, well thought-out, and relevant" (I/2nd). And a student said: "The teachers want to teach me to manage on my own" (I/2nd).

By working independently, students were encouraged to find their own pace, to face and overcome frustrating situations, and to develop a system of internal control, a personal discipline and style. Another professor said:

> Students must learn to see whether the experiment is going well and to correct it [if necessary]. Although we give them formulas for how to do it, it's something that they have to learn on their own and through experience.
> The student is left on his own to research the techniques he will need. . . . At the beginning [I let him] do it however he can; . . . later I guide him so he will see which techniques work. . . . The student follows a minimal amount of instructions and discovers from the beginning how best to choose his methods. (I/2nd)

A difficult experience, but one that is important as a socialization mechanism, is to let students work on their own. Students learn and live objective as well as subjective elements of the role of researcher; they learn to take initiative, to look for information, to make decisions, and to develop strategies for solving problems, for organizing their time, for being responsible, and for managing on their own. One second-semester student said: "I felt that I was progressing very slowly and that it would have been faster if they had given me the technique [instead of making me go look for it] . . . but that doesn't happen in the life of a researcher" (I/2nd).

In the second year, the professors integrated the students into a research group. This helped the students to see themselves as part of the research community, since the entire laboratory worked on solving the same problem. Nonetheless, each student was to be responsible for her or his own work, which made the students feel that they were gaining recognition as researchers. The professors would encourage the students themselves to propose (or to think they were proposing) the experiments. Noted one professor: "[In the discussion,] experiments came up which we had led them to propose" (I/3rd). The responsibility of each student having to work on a problem intimidated the students, although at the same time it made them feel mature: "The distressing part was that for the first time we began to do something on our own. . . . Even though each one of us had a teacher/adviser, it was no longer the same, because he would tell you the general technique but not the details" (III/3rd). In this manner, independence, self-sufficiency, and individuality were stimulated, adding to the students' awareness of the intense peer pressure.

We drew no distinctions between the first and the third cohorts because we found that the dynamics of the socialization process were similar. Starting with the third cohort, however, the professors insisted even more that students work on their own, and that they choose not only the experiments but the problem to be worked on. The professors said:

> If we continue with the premise that [we should not] spoil them with intensive participation by the professor . . . [in other words, that] the professor should curtail his participation . . . [the students] might mature well. (I/3rd)
>
> I tried to make each one of them draw his own conclusions, and to let the others discuss and criticize them. Sometimes it's

good for everyone to arrive at his own solution, but it's not possible to stop them from communicating with each other. I insisted that each one work alone, and in the end they did. This took a certain amount of time. (I/4th)

The process of becoming independent is gradual. During the first year, the students are told to look for information and to learn techniques on their own; they are also allowed to discuss and propose experimental ways of solving problems, although they are given guidance. Thereafter (in the second year), each student conducts her or his own experiment, so that she or he will eventually be able to propose a research problem; this culminates with the thesis, at the end of the undergraduate program.

Allowing students to work on their own leads them to feel responsible and to push themselves to work, provided this mechanism is buttressed with supervision by the teacher/adviser. Some professors were aware of the importance of maintaining a balance between guiding the students and allowing the students to work alone:

> The idea behind some teachers' not giving the students details of the techniques to be used is to stimulate student creativity by making them look for the methodologies; however, I believe you have to keep a commitment. . . . [I]t's all right to pose the problem to the student and let him postulate solutions to solve [it]. . . . This is important for creativity, but I think that when the experiment per se is being designed, that is, once the strategy has been decided, in accord with the professor . . . [the student] should be provided with the details of the methodology. (I/4th)

When the balance between guiding students and letting them work on their own is upset, students feel abandoned; they stop working, or they work little and poorly. This occurred, for example, in the fourth semester of the first cohort, when professors were absent or hard to find. Students felt that they lacked direction and lost their motivation:

> I started to read and read. [The professor's absence] didn't affect me, but I began to get bored; then I began thinking and I realized that I was progressing slowly because, instead of that, I could have done an experiment on my own. (I/4th)
>
> When I was unable to find the teacher/advisers . . . the work

seemed slow; . . . the laboratory work seemed mechanical, and I lost interest. (I/4th)

As might be expected, the least dependent students were affected the least. We found one student who, from the beginning of the second year of the program, attempted to work alone, unsupervised. This was the only such case in the cohorts we studied.

Supervision and Guidance

In reality, students were not as alone as they were made to feel. The professors followed their movements closely, even though they let them work on their own. They would visit the students, watch them work, pay attention to their behavior during the discussions and in the laboratory, and ask them about their work progress. At times, professors would guide the students directly, telling them what to do and how to do it, commenting on and discussing with them the progress of their work. On other occasions, their guidance was more subtle and occurred in the discussions and questions that led students to make decisions.

The role of supervision in the socialization of the students is of fundamental importance, since it introduces a personal, warm element into the professor–student relationship: "I've read a lot, but when I left the classes in which there was more guidance, like [the ones] I took last semester, I would leave very excited. The topic of the course [I'm taking now] fascinates me, but it doesn't allow me to get as excited" (I/4th). Supervision was the foundation of the relationship between the teacher/advisers and the students. The professors would guide students continuously and pay attention to every detail: how the students handled information in presentations, what concepts they used, what sort of questions they asked, their personality traits and how those traits could be channeled positively into their roles as scientists.

The second year shows clearly the relationship that the professors established as guides and supporters of the students. Guidance was given constantly, although it was generally indirect and concealed behind an apparent total freedom to think, discuss, and propose. Thus professors would provoke discussions and lead the students to ideas and experiments that had been planned in advance. One student recognized this tendency: "We are supposed to participate in the planning of the experiments, but in the end the one that [the teachers] have planned is chosen" (I/3rd). Nevertheless, this was not a rigid mechanism, and professors were open to accepting original ideas; on occasion the students would propose experiments that modified the ideas or experiments which the professors had proposed for development. One professor noted: "An important part of the course was that we had very stimulating discussions. I was enthused to see how the students were able to tie up loose ends. This is a good group if you orient it and ask it the right questions" (I/4th).

The type of guidance each professor provided depended on his or her personal style. Some professors let the students work on their own and gave rather general guidance, whereas others led the students more closely; some pointed students to general ideas and experiments but did not help them specifically with the practical work, whereas others directed the course of the experiments, providing methodological details and information.

Not only did the professors supervise and guide students in their work, they also concerned themselves with the students' moods and were quick to detect, for example, depressive states. One student said: "I felt bad because nothing was working out; I ran into [the professor] and he told me he had noticed that I was depressed because my experiment wasn't working, but that I shouldn't feel like that—that things always happen that way. I don't feel so bad now" (I/4th). A professor commented: "I stress that experiments don't always work out, but that they shouldn't feel depressed or dejected; on the other hand, when their experiments work you have to calm them down a little" (I/3rd).

Teachers would offer this kind of support in their frequent, personal talks with students, in which they always tried to appear as strong figures, sometimes in an exaggerated way, even though they felt quite differently. One second-year professor said:

> Sometimes I have to spend my time consoling people because they got a piece of data they didn't want; sometimes I have to adopt an attitude of raising their morale, of telling them that their work and mine are valuable, and sometimes this is difficult. . . . Many times I have had to hide my true feelings and I have tried to present myself differently from the way I feel . . . to adopt an attitude that, when you say it, sounds positive, like one, for

> example, of self-appraisal, though one is [in fact] far from having [this attitude]. This is a pleasant image, but [it is] not complete [or] honest. It would be much better for them to know ... what a researcher truly feels regarding the appraisal of one's own work. (III/3rd)

Part of the supervisory work had to do with gradually fine-tuning the students, both regarding the assimilation of the values being transmitted to them and regarding modes of behavior. Thus, they were told to be critical and imaginative, that the information was less important than their training. Nevertheless, the professors began to evaluate students when they went too far by, for example, being overly critical, putting forth grandiose ideas, proposing research projects that were too ambitious, or deriding laboratory data or work. It was common for students to propose overly complex, overly ambitious experiments; the professors would guide the students until they came up with more realistic proposals and smaller, more feasible experiments: "If the students proposed something that was out of focus, I would guide them a little so that the whole experiment wouldn't get off track" (I/4th).

The students did not clearly perceive their professors' protective attitude; however, they did notice and resent their aggressive attitude, even though at the end they appreciated it. Several teachers would—among other tactics—resort to scolding and pressuring their students, expecting them to be more assertive in the way they proposed and defended their ideas:

> I like very much the way Dr. [X] teaches because it seems like he's scolding you. At the beginning this inhibited me, but when I realized that [he did it] to stimulate us, I too answered ... always [maintaining] a good attitude. The professor attacks you constantly and makes you think; ... you begin to read more. You feel nervous when he's there, although at the same time he motivates you incredibly.
>
> The first time he yelled at me I was paralyzed, but I began reading. You feel that you're being attacked and you try to defend yourself and you study in order to do so. The discussion becomes your own. (I/3rd)

Evaluation and Rewards

Although there were no exams, the professors made constant evaluations. In the first year the pressure was relentless, and little praise was given. The students felt they were being constantly observed: "To feel the eyes of the Institute on these 'dummies' was terrible" (I/2nd). Or, as one professor stated: "There are five professors to watch four students" (I/2nd).

The professors would comment on the quality of the students' questions and answers. Some students asked questions on the material they had read; others simply pointed out the information they had not mastered or had grasped poorly. The teachers would criticize the students directly: they would point out their shortcomings; and they would tell them how badly they were doing things, without showing them the right way.

The professors evaluated the students according to their own ideal of what a researcher should be, and they conveyed to them clearly and openly where they felt the students were lacking and what they needed to work on. They did not, however, tell them how to do it. By contrast, the expressions of satisfaction and the incentives were very scarce. The professors did not reward the students' behavior, nor did they compliment them openly: "[The professors] say we're slacking off, that we're not doing things as well or as efficiently; at other times they say that we produce little, but at no time do they say, 'Wow, you've really learned a lot.' The teachers complain about our low output; well, so they should do something, or do they think we are lazy?" (I/1st).

The rewards were subtle and implicit: the professors conveyed signs of satisfaction in the discussions by giving students more opportunities to speak or by criticizing them less intensely; on other occasions, they would show approval by acknowledging that a student would make a good researcher in the future. Nevertheless, in the first year, students did not perceive these signs of approval by the professors: "Sometimes I would like them to tell me that I did something right . . . and [I would like] to not always feel that I have been mediocre" (I/2nd).

As the professors evaluated the students constantly and let them work on their own, they made them resist to a greater degree the high demands; the students felt forsaken and, at times, helpless. They sought, at all costs, to discover what type of behavior would please their professors, and they felt frustrated at not being able to find the answer.

Only at the end of the first semester did the professors give them a small sign of approval, telling them that "they weren't doing that badly" and conveying to them that they had overcome that stage of the program. Students said:

> They never told us whether we were doing well or badly. I think that it's useful for them to tell you how you're doing. Only now, at the end of the course, did they tell us we weren't doing that badly. (I/1st)
>
> I had the impression that there was no way for me to know what direction to go in. (I/2nd)
>
> We asked Dr. [X] to evaluate us because we didn't know where we stood; . . . we wanted to feel more secure . . . because we felt that we were bad [students], that we weren't making it . . . that the professors expected more from us. . . . The professor said that he saw progress in us and that he realized that they, the professors, had demanded too much. (I/2nd)

One professor said:

> We found that the students were frustrated because we refused to direct them as [is done] in traditional schools. . . . [At the end of the semester] we talked with them so they wouldn't be so down. (I/2nd)

The professors, then, did convey signs of approval, although the students did not always perceive them. In discussions, for example, the professors would attempt to instill confidence by making the students feel capable of participating, even if they didn't have a grasp on very much information. One student said: "The good thing about the professor was that we, too, were able to give an explanation of the phenomena, which could even be different from the [explanations] in the book" (III/2nd).

Evaluation and criticism stimulated the students to work harder by making them try to find a way to please their professors, especially when they felt their professors were personally concerned: "Personally, it bothers me to make people feel bad, [that] someone feels frustrated because of me, and that's why I feel bad when I feel pressured by the professors; I'm afraid to disappoint them." Evaluations also indicate to the students certain elements of the ideal to which they should aspire. They constitute one of the mechanisms by which scientific ideology is most openly and clearly transmitted—what to be like and what not to be like. As students internalize the model, they feel satisfied. One student, at the end of the first year, seemed to have accepted the rules of the game: "It's not very important to me that the professors tell me how I'm doing, because I feel that by solving a problem I'm doing well" (I/2nd). The same student had complained of the opposite problem when he entered the program.

In the second year the professors, though they continued to be critical and demanding, showed greater approval and confidence in the students, as can be seen in their comments on both cohorts:

> This is the best group of students that I've had. In the end, they would propose the experiments to me. They would have been able to continue doing research. (I/3rd)
>
> They question everything; it's like talking with another researcher. (III/3rd)

The professors' expressions of approval gave the students confidence and made them feel they were capable of proposing and conducting a good research project. Presenting and discussing projects in group seminars and in the scientific congresses they attended also gave them a sense of recognition and, therefore, satisfaction. And the professors tried to get their students to like their work, so that they would feel rewarded: "We had the students give their own names to the bacteria that they isolated. This made them feel very good" (III/4th). Nevertheless, the professors continued to pay close attention to all erroneous information and bad techniques, reminding the students that they were not yet the great researchers that they might think they were.

Conclusions

Through the close relationship they had with their students, professors came across as role models and transmitted the scientific ideology. Students saw them at work and in discussions. Based on their relationship with the professors in class, on what their professors told them, and on the criticisms they made, the students began to distinguish between their professors' different styles, identify some of their traits, and idealize them in a positive and negative way, as a step in the identification process. They idealized in a positive way such traits as intelligence, clarity of thought, imagination, wisdom, enthusiasm, and the capacity to transmit these qualities (which, moreover, are stressed in scientific ideology). It was possible for us (the authors) to know which professor the students were working with at the time we interviewed them. "[M] is a very good professor; he knows a lot. [X] seems to me to be a very good buddy and teacher because he makes you think; he discusses things with you until you are completely sure. At times, [P] inspires me to want to study and to go to my next class; at other times, he disconcerts me because of the way he is, which is very inconsistent. Sometimes he's too euphoric, and that's bad; it's all right to be euphoric, but [you shouldn't] go overboard" (I/2nd).

At times students expected the teachers to act in complete accord with the traits they idealized, and they would show intolerance toward any sign of weakness. To a certain extent, this reaction was reinforced by the professors' high demands. The students also criticized researchers for becoming "alienated" by devoting themselves completely to their research work. At the beginning, the students were afraid that they would become too engrossed, that they would lose their individuality by becoming completely devoted to research.

The scientific training included treating students as young colleagues; this implied not only raising them to the level of researchers, but also letting them see (since they were close to the work) researchers' weaknesses: researchers do not start off with a blank slate; nor do they follow the scientific method, or the ideal model, to the letter; nor are they perfect. This contradiction between the idealized image the students had of their professors, the image that the professors tried to convey, and the professors' actual behavior produced a feeling of disappointment in the students. This disappointment was gradually overcome during the process of maturation. At the end of the first year, the idealized image of researchers, by which good and bad aspects were dissociated (researchers were seen either as heroes or as social misfits), began to be integrated. Hence, one student said: "I see that researchers are not alienated persons, as I thought [they were], but [that they are] cultivated and know what they want; . . . they are persons who are anguished and constantly changing" (III/2nd).

By the second year, the professors had begun to establish personal relationships with the students and had begun to talk to them about their own research projects, about the way in which they worked and, at times, about certain details of their private lives. The professors were aware that they were acting as role models, and in general they were careful concerning their image: "Students are malleable, and part of their professor's style rubs off on them; in this way they gradually mature" (I/3rd).

We made the following observation during a Monday seminar (I/3rd). A professor was speaking with his students, asking them about their activities over the previous weekend. The students replied that they had tried to forget about the laboratory and did other things they liked. The professor said: "I took twenty books home and tried to solve a problem. That's not what I do every weekend—not because I wouldn't like to, but because of other obligations." He went on to speak of other interests: "It's like when children go to a toystore; they want all the toys, but eventually they choose one." One student replied: "My creativity can be expressed in other ways, such as photography, music, etc." Even in protesting, then, the student had imitated his professor's vocabulary and emphasis on creativity: they shared a vocabulary and part of an ideology.

Professors generally presented themselves as being strong, sure of their scientific careers, and in love with their work, which they portrayed as having great importance. They created an atmosphere of enthusiasm, which in turn established a bond between them and the students, by making the students feel that they shared in common a heroic task.

In sum, through the close relationship that developed between the teacher/advisers and the students, the professors became models for identification, for transmission of ideology, all while molding the students via the feedback that their behavior and ideas elicited. In addition, the students, by being in contact with a research group, began to identify with other role models, such as their laboratory companions, researchers from other institutes whom they met at scientific congresses, and scientists with whom they became familiar solely through their published work.

Interaction with the Group

Reference Groups

In addition to having a close relationship with the teacher/adviser, the student comes to form part of a group (see Table 10). At the beginning,

Table 10. Students' reference groups

Stages	Integration Group	Identification Group
First First and second years	Students in the same cohort at the IIBM	Teacher/Advisers
Second Third and fourth years	Laboratory coworkers	Identification with the group that works in the laboratory, its respective teacher/advisers, and with the Institute

Third
  Fourth Year
  Master's
  Doctorate
	Networks based on common interests	Identification with groups and persons recognized within the national and international scientific community; rebelliousness against teacher/adviser figures

Final objective	Networks and abstract community	Identification with disciplines and with the scientific community; acceptance of an abstract ideal

the group is made up of the student's other classmates and is led by the teacher/advisers. This is a temporary group; the student is not encouraged to identify with it. Starting in the second year, the student joins a work group (the laboratory), where she or he interacts informally with the head teacher/adviser, with other classmates, and with the technicians and researchers who make up the group. The discussion seminars represent the forums for formal interaction within the group. The student begins to identify with both the teacher/adviser and the group. Later, as the student begins to identify with a problem of scientific field, her or his group is extended to "invisible networks" of scientists who have similar interests (see Crane 1972). The interaction occurs, at that point, through readings, publications, symposiums, and congresses.

Discussions

The discussions were one of the main means of socialization. The discussions or seminars condensed a large part of the intellectual, emotional, and social elements that the Institute hoped to instill. One or more professors, in addition to the students, participated; hence, the discussions and seminars represented experiential situations in which students saw the teachers interact—and think—and in which they could imitate them. At the same time, the professors would correct and mold the students.

Discussion seminars are a social arena in which different thought functions are acted out. One participant formulates ideas; another questions them; others recall. This is one of the ways in which students learn to think. In the discussions, some students alternate between different positions and mental activities while other students watch; they gradually develop their own style and identity (Mead 1934).

In the first years, discussions introduce students to the modes of coexistence that are expected in researchers; the students are taught to be members of a group. As noted above, the professors would require the students to take a position regarding ideas that had been presented, and each student would defend her or his ideas and criticize others'. Discussions became a sort of war of the minds, in which students attacked each other's ideas; each student had to defend her- or himself. This would sometimes produce emotionally charged situations.

The teachers slowly led their students to present clear ideas, to defend themselves, and to criticize by using logical arguments, rather than emotions. Regarding this, one professor said: "Discussions with them are tiresome; you have to be patient, to show logically that the argumentations are correct, without cutting [the students] off in an authoritarian way" (I/3rd). One student noted: "I learned to use all the terms properly, not to intevene when we don't know what we're talking about; especially after the seminar given by [Dr. X], in which we were talking about mutation, and he said to me; 'Do you know what mutation is?' I said that I did, more or less. He repeated: 'Either you know, or you give the floor to someone else' " (I/3rd).

Just as professors made the student defend and attack their own and others' ideas aggressively, they also made them face the idea that different interpretations exist regarding any given fact. One student said: "I learned to think and to let [others] think, [i.e.] . . . to be open; . . . I can say, "I believe this," all the while leaving room for the 'but' " (I/3rd).

Moreover, the students had the opportunity to observe how their professors presented and defended their own ideas and criticized others' ideas. When several professors participated in the seminars, differences of opinion surfaced and were discussed. Through this, the students could appreciate different forms of reasoning and learn different styles of discussing and proposing ideas. In addition, they began to realize that every professor was criticizable—that she or he was not infallible—which reduced the hierarchy between professors and students, allowing disagreements to be openly expressed.

The discussions generally enthused the students, who quickly learned their own style and caught on to the spirit of discussion, even though they overacted this role. One student said: "I really like discussing [things]. . . . [I]t's what I like most; you learn twice as much—[you learn] what you think and what the other person thinks" (III/1st). A professor said: "At the end of the semester the students learned how to discuss, and now they are very careful about what they say" (I/1st). Another student added: "I really like the way [the professor] leads the discussions: he makes us think; he leads you to draw [conclusions]; he doesn't say anything even when sometimes [all of us] are shouting and arguing, because we really get into it, but it's neat and you learn. We got used to talking in a certain logical order, to developing everything, to explaining why each step was taken" (III/1st).

In summary: discussions teach socially acceptable behavior, the rules of the game for academic debate (how to discuss, attack, and defend ideas; who wins and who loses; etc.). Discussions represent the basic unit of social life in the scientific community. On the one hand, this is an essential form of "community," in which individuals who are struggling to produce and impose their ideas confront each other; on the other hand, it provides a feeling of communion with a thinking community. It also gives one the satisfaction of taking part in the activity of thinking; of seeing one's own mind—and the minds of others—at work; of listening to the words of others as a response to one's own words; and of gradually assuming a role within a social group. Scientific discussions have their own "rules of etiquette": one must know how to attack and be aggressive in using works, and one must always remain within intellectual and scientific—rather than emotional or personal—boundaries. Likewise, one must tolerate criticism and be able to assimilate it. Finally, through discussion seminars, each individual gradually establishes her or his style and status within the group.

In discussions, then, students learn the ways in which they are to relate horizontally (among equals) and vertically (with their professors), as well as the rules for both working together and competing. Moreover, discussions give students public recognition, making them feel capable of having good ideas and of defending them well. This is one of the few immediate rewards students receive in the early years of their training. The discussion seminar is part of the scientific community's secular ritual (congresses, symposiums, formal seminars, conferences), in which there is an exchange of ideas and knowledge and in which social recognition is given.

The Work Group

Belonging to a generational group was not particularly stressed in this undergraduate program. During their first year, students were together as a generational group; nevertheless, they were encouraged to develop their individuality and competitiveness, to learn the rules of coexistence, and to establish links with the professors in order to achieve a social identity with the larger group (the laboratory, the Institute, the scientific community). These links with teacher/advisers represented, in the second year, the foundation for identification with a work group, led by a chief researcher, who was usually the seminar professor. One professor noted: "[The students] become part of the laboratory. They live here; they are like our coworkers; we don't segregate, and that's important" (III/3rd). Or, according to one student: "The most important thing about this course is that we went to work with a research group, and not with a professor as [we had done] before. Here the idea was to see how research is done in the real world; I found out what it is like" (III/3rd).

This social milieu gave the students a feeling of camaraderie and confidence, as well as a sense of belonging; this was gratifying to them, for they perceived validation and recognition in it, and they considered themselves to be young researchers who belonged to a research group. Students said:

> The semester began in March, and now in April I feel as if I had been in the laboratory for two years. (III/3rd)
> It is easier for me to integrate with the people in the laboratory; ... they also seem to be interested in helping me. (I/3rd)

The transition to becoming a member of a work group represented the student's true social initiation: this was the social unit of the scientific-experimental community. These students learned to work alongside more advanced (master's and doctoral) students as well as with students who were behind them, with other researchers in addition to their teacher/adviser, with technicians and, to a lesser degree, with the students in their own cohort. This taught them a set of rules for social behavior and indicated to them what the structure of a work group was.

Symposiums and Scientific Congresses

Another important mechanism for socializing students and transmitting ideology is attendance at symposiums and scientific congresses. The congresses allow students to meet and compare different researchers, to see what is behind a research project, to appreciate different styles of presenting projects and observe how those projects are criticized and defended. Moreover, the students can see how recognized scientists act and interact; these scientists may become role models. One second-year student said:

> One man, who has been nominated for the Nobel [Prize]—a worldwide authority in genetics—and who is someone you could understand . . . he's on the cutting edge of knowledge and, after working on one problem, his experiments are working; and that could be seen. . . . He explained how he had thought of the experiments, and we could almost see him planning and thinking; you could tell that he knew [what he was doing]. We left there fascinated and motivated. . . . In addition, he made me think of the implications that what he is doing could have, and I haven't seen that in any other professor. You realize how important it can be when an experiment works and what implications it can have. (I/4th)

Another benefit from the congresses is that students consider themselves to be part of a work group vis-à-vis other groups; they develop a feeling of belonging to a particular scientific group which, in turn, belongs to one of several different theoretical factions that confront each other in the congresses. Among the student comments in this regard were the following:

> At the congress I realized that group research is very important; you must work [with others] in a group or an organization, because what you do is helpful to others and vice versa. (III/2nd)
>
> I realized that the Institute is one of the places where the best research is done in Mexico. (I/4th)

Gradually this also encourages students to identify with a larger community: the community of a specific discipline. By meeting researchers from other parts of Mexico and the world—researchers who work in the same field—students' concept of community is broadened. One student said: "The Congress on Nucleic Acids in Cuernavaca was very interesting. . . . [I]t allowed us to become familiar with the scientific community, to see what others do, because before we only knew the Institute's [researchers]" (I/4th).

At the congresses there was also personal contact between students and researchers. Students said:

> We swam and ate together, and it was neat to see the professors in another environment. (I/4th)
>
> One neat thing about the congress was that we dealt with the professors as real persons, because they are serious here, more natural there; we saw them on another level—swimming, dancing. (III/2nd)

In scientific congresses, other models are made known. Also, an idealized image is sometimes shattered: researchers dance, drink, get drunk, and in so doing are temporarily released from certain controls; in addition, hierarchies are relaxed. In conversing and arguing with researchers having a higher rank—even just making themselves heard—students feel they receive recognition, which can produce in them a rush of euphoria (Lomnitz 1981).

Finally, the plenary sessions of the congressses are rituals in which an outstanding member orally transmits scientific ideology and/or an exemplary piece of work. The official awards that are made symbolize the qualities that the community values in a scientist and serve as an example for younger cohorts.

Starting in the second year of the undergraduate program, an internal congress was held at the Institute in which each student (in all cohorts) presented a poster explaining the experiment she or he had conducted.

Students, professors, and guests visited the students, who explained what they had done. Such congresses were true celebrations—fairs for exchanging and valorizing information. Students felt like researchers; they felt that they "owned" their work. The congresses ended with a meal celebrating the end of a year of achievement. In this way, students began to prepare themselves to present articles and to feel the recognition and satisfaction of doing so.

Role-Playing

From the outset, one of the objectives of the undergraduate program was to integrate students into scientific life and to encourage them to acquire an identity as scientific researchers through their daily work. Thus, from the first day, students carried out laboratory work; an effort was made to make them feel that they were acting like researchers as they solved experimental problems.

At the beginning, especially with the first cohort, role-playing took place mostly at a level of discourse that stressed the importance of training and creativity. Hence, students were told that researchers should not be encyclopedic, that they should not try to accumulate knowledge, but to create and apply it. Professors required that the students apply their knowledge to the problems at hand, and that they discuss these problems so that they might arrive, on their own, at the concepts being taught. This translated, initially, into the students identifying originality, creativity, and criticism as the determinants of a researcher, and into their trying to show these qualities in an exaggerated manner. Thus, in the first semester, little value was given in discourse to the importance of information; therefore, students did not identify this element as being important to their role. This caused some problems. As one professor pointed out: "The emphasis on being creative that they received in the first semester might have caused the students to focus too much on interpreting or creating information, rather than on acquiring it" (I/1st).

The students themselves also began to feel the need to acquire information:

When I discuss amino acids and their characteristics, for example, if I don't know a formula, I don't understand the discussion. Then I have to study the topic in order to understand or solve the problem. In this I find a contradiction with what the professor wants, because he says that you shouldn't study theory per se, but only in order to solve a problem. (I/3rd)

And another student described his experiences as follows:

We are studying mitosis, though the main objective is for us to orient ourselves in research so that we will see what information we need and where to look for it. . . . At the beginning we were lost, overwhelmed with work; but now we've more or less caught on, although we don't yet know how to interpret the information and discard what is not fundamental. . . . [A]nd the thing is, there was a misunderstanding; we felt that we had to know everything, but [the professor] asks you for a minimal amount of information. (I/2nd)

The acquisition of information—both theoretical and technical—thus became a part of the students' work. To solve any problem, they had to look for the information: "I have learned to look up what I need in books. Mainly, I am getting information in order to use it, [rather than] memorizing it" (I/2nd).

A conflict arose, especially in the early years, between the need to have certain information in order to carry out scientific work and the importance professors gave to students' initiative and independence of thought. Nevertheless, what the professors stressed most was what they called "student training," which had to do with learning ways of thinking, including raising interesting problems, designing strategies to solve them, and knowing how to communicate them (see Chapter 3, above). This was taught through reading assignments, discussion seminars, and personal dialogue between teacher/advisers and students.

Reading Assignments and Seminars

Reading assignments teach how other scientists reason, and they present formal work models—that is, research "the way it should be done"—by showing the final product in the form of an article. From these readings, the student extracts the models and work-styles that are to be imitated and learns to recognize the different formal steps in a research project: locating relevant information, formulating a working hypothesis, choosing a methodology, and obtaining the results.

Starting in the first year, the seminars for analyzing and criticizing reading assignments had a special importance. In these seminars, the speaker would outline the information given in an article, point out what was most relevant, and interpret it. In this manner, students developed the ability to communicate ideas. Later on, the professors would encourage students to comment on an article critically, or they would start things off by criticizing a theoretical or methodological point.

By being pressured and criticized in oral presentations, students were obliged to take a position regarding the article they were reviewing (its theory, methods, and results) and to support their position with logical arguments. Students soon understood the importance of having a critical attitude: "We don't know how to read or criticize the articles . . . as to their possible truth [or] reliability, nor do we know how to discuss theory. When I read, I swallow everything; . . . I believe it and I don't discuss it. Then it turns out that there are several problems that I hadn't thought of" (I/2nd). In addition, students were urged to doubt, question, and argue.

Some professors began their classes by raising questions, and they also demanded that the students in turn raise many of their own questions; they would commend the students publicly for those questions they felt were incisive. One professor said: "I try [to get] the students to ask many good questions . . . and by *good* I mean they should be clear, well thought-out, and relevant" (I/2nd). The message being conveyed was that thinking about and proposing ideas was at least as important as obtaining information, and that the purpose here was to stimulate creativity. In these discussions, moreover, it was stressed that scientific vocabulary should be employed rigorously: "We had to know the precise meaning of each word we used" (III/1st).

At the beginning it was difficult for the students to participate in discussion seminars, but the professors would insist that they do so: "The professor makes you think and say what you think; he argues with you until you are completely sure or until he makes mincemeat out of you; you have to support your position with logical points constantly. I would call his class 'Scientific Thinking' " (I/2nd). Most of the students suffered because of the level of aggression in the discussions, although they considered them important to their training:

> [That's] because you must base everything rationally, . . . and this is a pleasure for the mind. (III/1st)
>
> One thing that I like about the professor is his very special way of attacking you, because it makes you react like a lightning bolt; he leads you so that you'll think and give reasons that back you up. Sometimes the professor takes absurd positions so that you'll refute him, . . . and we have to find a way to criticize him. (I/2nd)

In the second year, more was demanded of the students, although the discussions were conducted in the same fashion as before: not only did students have to present a reading assignment in detail, but they also had to put forth their own ideas and the experimental results. The students would review classic texts and scientific articles, some of which had been written by their professors. The professors discussed everything with their students, making them state meticulously their ideas and those of others, without overlooking a single detail, regarding either theory or the technique being used: "Now I read differently; I care more about facts than about interpretations. When I have the data, I compare what I think with what the author thinks" (III/3rd). By this time, too, the students had begun to design their own experiments and to play the role of incipient researchers: "You could say that this is the first semester in which we began to learn research design. Last semester we had to count red corpuscles, but now there are established techniques, and the only thing we did was to choose from among two or three techniques; now, however, the idea is to let your imagination wander—the idea is to create different experiments and decide which one we think will work" (I/3rd).

In general, the students would propose grandiose ideas and experiments. The discussions served to focus their initiatives, making the students come down to earth until they arrived at proposals more in line with their knowledge and circumstances: "At the beginning we had very lofty goals, excessive imaginations about doing superexperiments, but the discussions with [the professor] brought us down to earth" (III/3rd).

The following comment by one teacher exemplifies the professorial role in the discussions:

> Before we give them the experiment, we the professors who teach the seminar agree on what questions to ask and how to ask them. Sometimes we have two-hour discussions [with the students] on a question. . . . At the time of the discussion, we tend to make it as open as possible, [letting] the students propose whatever comes to mind, and as the discussion advances, what is useless is discarded until the students have formulated the experiment themselves. . . . As the experiments are carried out and observations and comparisons emerge, we lead the students to propose new experiments in the discussion. (I/3rd)

We were able to observe that the professors took seriously what the students said, which gave the students self-confidence. The professors would participate in the discussions, proposing their ideas and commenting on their own research projects. By revealing their way of thinking and working, they came across as role models.

We have again analyzed student participation in the discussions because of the fundamental role of communication in a researcher's life. Thus, role-playing as we have described it includes not only the students' laboratory work but their participation in the exchange of ideas as well. The elements of research work that are learned through the discussions include 1) a scientific vocabulary and the concepts it covers; 2) the way in which information and ideas are presented, analyzed and interpreted, criticized and defended; and 3) the manner in which questions are asked. In other words, students learn to think.

The discussions transmit such messages as these: students are able to think on their own and can criticize everyone and everything; students have the same potential as their professors; one must see things from different perspectives, have many ideas, be creative; it is all right to disagree with an opinion held by others; it is important to question everything that is written and spoken; there are no absolute truths; everything is fallible, and everything can be improved; there is not merely one interpretation of facts; a researcher's abilities are constantly being tested. Likewise, we noticed the transmission of contradictory messages—messages that reflect different aspects of a student's life. For example, students are told to think logically and to be careful in their judgments, but at the same time they are encouraged to look for "wild ideas" and to discuss ideas without even knowing how.

Laboratory Work

The laboratory is the experimental scientist's natural habitat. Students spend most of their time there and begin to feel that they are in their element: the odors, furniture and instruments, the white robe, etc. make them feel like scientists. This, however, requires a learning process that is not always easy.

From the beginning of the undergraduate program, students were subjected to an intense work routine. During the first year, the class schedule was from 8:00 A.M. to 7:00 P.M., with a break from 2:00 to 4:00 P.M. The work schedule was very tight, as they took twelve hours per week of theoretical classes, eight of seminars, and another six of laboratory. They also devoted about ten hours to individual study and laboratory work. Hence, the students remained at school all day, which caused many complaints, since they were accustomed to spending half their day at home: "I produce less by being here all day. I prefer to go home after 2:00 P.M. . . . I'm fed up with being here so many hours; I like to study however and whenever I want" (I/1st).

In the second semester students were given greater freedom to manage their time according to their own needs. There were no fixed, obligatory schedules, and students could leave the Institute at any time; nevertheless, it was observed that they preferred to remain there, not only until 7:00 P.M. but sometimes until 10:00 P.M. They would also go to the Institute on weekends and during vacation periods. This was necessary because of the intense pressure of the academic demands and the type of experimental work they carried out, which included slow operations at irregular intervals. During the first year, students complained a lot about what they felt were excessive demands, which forced them to sacrifice any interests not related to research. Still, they became accustomed to working long hours and to devoting whatever amount of time was necessary for the progress of their experiments, regardless of the time of day.

Starting in the second year, the schedule was much more flexible: except for six or eight hours per week of classes, seminars, and talks, the students' time was divided according to their own needs. It was the students' responsibility to manage their time and adjust their work pace so that they would be able to meet the goals set by their professors. Many students gave up their hobbies and other extracurricular activities because of a lack of time.

After the first year, it was observed that the students had acquired new work habits, that they had become tenacious and dedicated, and that they had a more positive attitude toward intense work. When the work pace slowed down, as it did in the fourth semester of the first cohort, students missed the pressure: "We had so much freedom that we worked very little; we did almost nothing. I went back to my old hobbies and I read a lot. At the beginning I thought this was good, but later I was exasperated because I had nothing to do" (II/4th). At this stage, students still needed to be pressured by their professors in order to maintain an intense work pace, even if they had learned to consider such a pace necessary.

By the final year of the undergraduate program, when many students were working on their thesis topic, the work pace had been sufficiently assimilated and was no longer questioned. The problem of knowing how to organize oneself and how to manage one's time persisted, but by then this was considered a definitive and permanent aspect of a researcher's problems. It was also observed that the third class, which was introduced to experimental projects much earlier, had fewer problems in accepting the intense work schedule.

When they were playing the role of scientist, conducting experiments allowed the students to appreciate what is behind an article or research project. They gradually recognized the difficulties that emerge in research and the importance of details.

In the first year, professors left the students alone in the laboratory, so that they could discover techniques and apply them. At times, the professors would correct or explain simple techniques or tell the students where to obtain information on these techniques. Laboratory experience was a comedy of errors that the students had to overcome. They were confronted with the need to carry out each step carefully in order to detect and correct their errors (see Chapter 3, above). The advice given by the professors was incidental and a posteriori. Students would comment on their problems and analyze, retrospectively, the steps they had taken, until they found the mistake; professors would almost always encourage students to find the mistake themselves; and although they would frequently give the students clues about what was occurring, the professors would not give them all the information. Students would repeat the entire experiment several times, since they generally would not find the mistake until the end. One student said:

> You don't think beforehand about what the experiment is going to be like; rather, in doing it, you see the complications.
>
> We saw how things are really done. We had to do the experiments alone; we had to have them ready within a certain amount of time. . . . All of a sudden, we would drop things; [the experiment] would be ruined because we had left certain things for too long or we didn't know how to prepare the reagents correctly. I had to learn to put in one chemical [first], and then the next one. (III/1st)

Each student worked at his or her own pace and, together with his or her classmates, began to become familiar with what it means to experiment. The students had to face the slowness of the experiment, observe the most minute details, detect their mistakes, and repeat the experiment over and over until they obtained the results. This aspect of role-playing forced them to confront the frustrations inherent to scientific work by showing them how difficult it is to do research. It was a laborious process, and the students supported each other (unlike their teachers, who gave them support as if with an eyedropper). One student said:

> I am very disappointed because nothing has come out [right]; I have done it three times. Nothing has come out [right] for anyone. Now I'm going to try the experiment with other chemicals. (III/1st)

Other students, evaluating a project that had lasted one month, pointed out:

> At those times, I felt that I was not progressing . . . that I was doing it too slowly. I think it would have gone faster if they had given me the technique, but I know this doesn't always happen to a researcher, so I think it was good to learn to go slowly. (I/2nd)
>
> We had to repeat the experiments a thousand times, so I now feel that I know how to work in the laboratory. (III/1st)

Still another student described the typical laboratory experience as follows:

> You want things to come out right, but as they don't you get depressed. I've been doing the experiment for two weeks, and it

> doesn't come out. The first two times, I didn't get anything; the third [time], my cultures became contaminated and were lost. In addition, it's difficult to observe mitosis because it's difficult to see the little sheets under the microscope, so I still don't know if the experiment came out or not. I'm desperate. (III/1st)

More importance was given to experimental work in the second year than in the first. The experiments were more complex; to solve a problem, students had to conduct sequential experiments. One student said: "We do an experiment from which we get a lot [of material for] discussion, from which more problems emerge, and [based on] these we do another experiment, from which we get [material for] more discussions, more problems, and more experiments. This has been very good" (III/3rd). This led the students to become meticulous—since any error had serious consequences—and to develop their analytical ability, as they had to foresee difficulties and think rigorously as well as to be patient and overcome frustration: "We have [to learn] to know what it is one wants to get from the experiment and, when it doesn't work, to explain why it didn't work" (I/3rd).

Doing more-complex experiments gave the students satisfaction, and made them feel that the laboratory was a world they knew and in which they could get along:

> I really like the way of learning here . . . like researchers, working with experiments, and not like students. (III/3rd)
>
> I'm very pleased. In the laboratory, you feel very good knowing how to do things. Now that we have experience, we know how to handle ourselves in the laboratory, we have more knowledge. I feel more comfortable. . . . I think that I'm getting more into all this. (I/3rd)

In sum, lab work was of central importance to the students' training. In the laboratory, students played the role of scientists—albeit incipient scientists. They learned to plan strategies, foresee events, observe, and compare and choose appropriate techniques. By doing their own experiments the students were compelled to develop a work discipline; that is, to organize their time, to be meticulous and orderly. Likewise, they developed a sense of responsibility and commitment regarding their work. The continual mistakes led them to develop patience and perseverance, to learn to tolerate frustration and distress. In the second year—as opposed to the first, in which thinking and discovering were stressed—professors spoke of the importance of efficient, careful laboratory work for testing ideas. One student noted: "We disdained the 'kitchen'; we thought that the important thing was to think, to formulate a good hypothesis; we didn't pay attention to how [you should] hold a tube.... [G]etting our hands dirty and doing things in the laboratory—that's also work, even though [you can't] forget about the theoretical aspect" (III/3rd).

Socialization in the Third and Fourth Years

The first two years represented an initial stage in the socialization process. Students had been exposed to scientific ideology through the work routine, through their relationship within a group of researchers, and via the discursive transmission of beliefs and values.

By the third year, the students had accepted the ideal model of what a scientist should be, and they identified with this model and with a group of undergraduate-level professors who had transmitted and represented it. This identification extended to the Institute. The students also acquired specific scientific knowledge and a set of basic techniques, accepted the work routine and the ideal of devoting themselves completely to their studies, and to a greater or lesser degree (depending on the cohort and the circumstances) learned to manage certain emotional aspects implicit in the role of researcher.

The professors felt that it was important, in the third and fourth years, to send students to different research centers in order for them to meet other researchers and become familiar with other styles and "families" that could broaden their interests and, ultimately, open up job possibilities for them. Likewise, it was hoped that the students would become familiar with research conditions in Mexico and with the country's needs so that as scientists they would be concerned with and committed to development and to solving the nation's problems. We should recall that a nationalist discourse was part of the earliest discussions on developing the undergraduate program, the purpose of which was to train researchers of very high caliber, committed to their country: researchers with an identity both as scientists and as Mexicans.

It turned out, though, that the nationalist element of ideology did not play an active role in student training and remained at the level of discourse, which spoke of "researchers' social awareness." In the first semester of the first cohort, for example, a seminar was given on Mexico's socioeconomic problems, but owing to a lack of interest it was suspended. Likewise, trips outside Mexico City and to various hospitals were considered unsatisfactory, because the students did not do research but instead engaged in relatively unsupervised fieldwork or clinical practice. The professors felt that at this stage of the student's education it was more important to train first-rate scientists than to "waste time" working in conditions that were not conducive to doing good work. Hence, students ended up learning the elements of discourse referring to a scientist's social awareness exclusively at an ideal level. For example:

> A researcher must be aware of reality and of the world in which he lives—and not feel that scientists live in a separate world. (I/5th)
>
> A researcher must not become alienated . . . or do [basic or applied] research that is not applicable to Mexico. (III/5th)

These fine words notwithstanding, the reality was that they did not choose to work in research centers outside Mexico City, nor were they attracted to research problems based solely on their importance for the country.

The professors continued to speak of the importance of making science relevant to the country's conditions, although they also emphasized the need to impart good scientific training: "We want to train very good researchers who can work on scientific problems related to solving national problems" (III/5th). From this point on, an effort was made to maintain this new line, not only in the undergraduate program but in the Institute generally.

The contact with external groups during the first generation's first year showed that students had identified with their group and that they had internalized many of the values taught at the Institute: "The professor [who was from another institution] did not quite understand what we, as research students, were doing. I had not had the experience of seeing a person who tried to be like a researcher without actually being one, as was his case, [and] who was not critical; all of a sudden he has a brainstorm and publishes everything without giving sufficient proof, [and] he changes the colorant and the technique according to whim. Only when you live through that can you see how important our degree program is" (I/5th). The students compared both groups and made note of any deviation from the work method to which they were accustomed, thereby reaffirming their sense of belonging to the Institute and their desire to be recognized by their fellow researchers.

One of the keenly felt problems in these off-campus experiences was the lack of closeness to the professors: "That was a very bad course, because they supposedly do research in it, but [in fact] they don't do anything. I kept my distance from the professor but he never paid attention to me, and the few times he did I ran circles around him. . . . In the other hospital, the course bored me because, more than anything else, I did whatever I pleased. No one paid attention to me anywhere" (I/5th). Traveling away from the Institute and interacting with teacher/advisers and groups that represented distinct research concepts, different mystiques, and a more distant treatment of students encouraged the IIBM students to play the researcher's role as they had learned it at the Institute; however, when they did not receive positive responses from their professors, there was a feeling of rejection and they wanted to return to the Institute.

Nor was the attitude of the outside professors marked by an excessive fondness for these undergraduates: "The students' relationship with me was one of competition. They didn't see anything in me they could feel envious of. They are from another family. . . . I saw them as spoiled children from an excellent home" (I/5th). Thus, the first cohort suffered during the third year from a lack of rapport with teacher/advisers and from integration into a mediocre work group, which affected their output and motivation. When the students returned, the undergraduate professors complained about how little work they had done and about the poor quality of their performance.

By contrast, the third generation's outside experiences were limited to research groups that had the same work-style as the Institute's teacher/advisers, for which reason this cohort continued to be exposed, uninterruptedly, to the original socializing models and could make progress with fewer hindrances. The importance of a close relationship with the teacher/adviser, group identification, and work itself in the process of socialization would become apparent.

Although the relationship with teacher/advisers did not lose importance, the relationship with the group eventually assumed greater importance as the object of identification, emotional support, and work: "I like the group's philosophy: they work in a very integrated manner; everyone takes interest in your problems. You feel that what you find interests them and that your data are part of something which is useful to them" (III/5th). The IIBM group, then, represented a good source of motivation and feedback: "It makes you feel you are doing something that has value; it sustains you, stimulates you, pushes you forward" (III/5th).

The group constituted a complement to and even a substitute for the teacher/adviser: "A researcher can't do anything alone—without the people who work with their hands in the laboratory; . . . the group teaches you the techniques; with [the group] you can talk more than you can with your teacher/adviser concerning the technical problems related to your project" (III/5th). Indeed, at this stage students participated as productive members of a work team and began to be recognized as young researchers. They conducted research, following each required step, even though they did not yet work on their own problems, but on their teacher/advisers' problems, which in fact were the work topic of the entire laboratory: "We all ran into [the problem] that our project had no purpose. At the beginning you have a goal, and it's very extensive; when you are working you realize that you aren't going to reach it or, similarly, you get things on the side that you had not even planned but that are interesting nonetheless. That's what research is like. In a way, that's bad, because you are never able to do what you said; but it's also good, because you follow other paths" (III/6th).

At the end of the third year, articles were published on these research projects. By now the students had assimilated the manner in which they were to work, and they were also aware that they did not always succeed: "We have an attitude of challenging or looking for alternatives to what someone says. . . . The important thing is not so much what you know, but that you consider alternatives and try to understand things. . . . [B]ut our shortcoming is that we can't accept that two ideas may be equally possible; we always try to have just one [idea] remaining, and in reality it's not always like that" (III/6th). We can see, too, how students identify with the work method that has been preferred at the Institute: "We don't like to read reviews that just give you data without explaining why. They don't describe the experiments; they don't explain how the data were obtained. You can't judge and confirm the conclusion, and that is the heart of research: you address a problem and evaluate the experiments; you have to know if the data respond to what was set forth" (III/5th).

The students also attended scientific congresses and presented projects. In the case of the first cohort these projects were assignments given by the immunology professor; nevertheless, the students presented and defended the projects, in the absence of their designers. In the case of the third cohort, students presented their own assignments. Such experiences gave them confidence, for they felt they were capable of presenting and defending a research project and that they were treated as researchers. In addition, they were satisfied that they had met with certain forms of interaction in the congresses and had been able to handle them: "I had to learn to defend myself against people who bombarded me with questions; . . . it gives you confidence to realize that you are knowledgeable and that you can answer [the questions] they ask you" (I/6th).

The participation of students in congresses, as members of the IIBM, reinforced their feeling of belonging to the Institute. By meeting members of other research groups, students strengthened their ties to the Institute and evaluated other researchers. This process culminated when the students, in conjunction with their teacher/advisers, published articles on projects they had presented at the congresses.

In conclusion, the different experiences that the first and third cohorts had during the year show the importance of establishing a close relationship with the teacher/adviser and the possibility of identifying with both a work group and a research project. Third-cohort students acquired more self-confidence than the first-cohort students. The professors also perceived an important difference in the progress the two groups made in their training, both intellectually and emotionally.

At the end of the third year, students from both cohorts had to begin thinking about a thesis topic, and this represented a difficult decision: "I feel awful, because I have to make a decision regarding my field, and it's very important because that's what you'll devote your whole life to. . . . I feel very ready—that I can do whatever they ask me to do. . . . I don't know very much, but [I do know] the basics and how to acquire the rest; yet, I want to test myself—to work on a project for which I have responsibility" (I/6th).

At the beginning of the fourth year, the students had to choose their thesis advisers. This was a two-way decision: students generally chose advisers who they felt were interested in them: "The words may be more or less the same, but you realize when you are really welcome" (I/8th). And the professors also set their sights on the students who they felt were most like themselves. Indeed, it was the professors who determined the selection; for if they did not show any interest in a student, they were in fact rejecting her or him.

In this final stage of the undergraduate program the students felt an even greater need to have a teacher/adviser:

> The idea for my research came from my teacher/adviser. Teacher/advisers would just hint at an idea and leave you thinking; they kind of subliminally give you the idea they want you to work on, and you think it's your own. That's what my teacher/adviser did. I think this is good because you don't have a clear idea of the field, of what's worth studying, although, in this way you also don't have the satisfaction of working on an idea of your own from the beginning. I hope someday to be able to work on an idea of my own. (III/7th)

> I had two important experiences with different teacher/advisers. One never teaches you; he speaks; when he's going to think about something he calls you to his office and thinks out loud, and this is how he teaches you. If you understood, fine; if not, too bad. I would ask him, "How do I follow this model?" He would say, "I don't know; you think about it, I can't tell you how to think." The other teacher/adviser teaches you to think; he gradually makes you develop your ideas and then his ideas, and he makes comparisons between the two, and in this way you learn to think with him. The second teacher/adviser's system works better for me, although I am more influenced by the ideas of the first. I see that it's important also to get along well with the teacher/adviser and over a long time: before, I had close but short-lived relationships; this is better. (III/8th)

> You always have setbacks—things that were done wrong—and there was the teacher/adviser, who helped me solve problems—big ones and small ones—at the moment [they arose]. I didn't feel "I've ruined the experiment"; I didn't feel distressed or nervous, as I did last year. Rather, I worked calmly; I felt very confident and that [I was] being taken care of, because at any time there was always someone to help me. (III/8th)

> I didn't like what I did, nor how I had to work with [Dr. X],

because you get little advice. In addition, since he has few people in his groups, there is no experimental foundation from which you can start in order to do a project; [it's not as if] you arrive and they tell you, "This has been done here, following these paths and these hypotheses." With [Dr. Y], who is working on a new topic, you have to start from zero, and at the beginning that's attractive, but it's also a problem because there's no foundation on which to begin. (III/8th)

I worked with [Dr. X] because I didn't like what [Dr. Y] was doing, because of the fact that he was going to do it alone; I have always tended not to have a very close relationship with the teacher/adviser. I see him little; a month might go by without me seeing him. I don't discuss my project with him much; I discuss things more with a master's student and his group. Sometimes I need to see him, and I look for him or for another teacher/adviser. I look for independent projects; I don't like to be supervised all day long; I need to discuss things, but I [don't want] them to tell me what to do. (III/8th)

In the fourth year, as in the previous year, students were members of their teacher/adviser's group and laboratory. All the students felt that they belonged to a distinguished group of researchers and that they worked differently from other groups, both inside and outside the university:

> I went to see laboratories in other [research] centers; people would tell me about their projects and I would raise doubts, criticize, and ask questions. (I/8th)
>
> I realized that [in that center] they weren't used to asking [questions] and discussing things. (III/7th)

By the fourth year, the students had an image of themselves as researchers:

> I have gotten more and more into research, and I see it as something that is very much a part of me.... I feel capable of discussing a problem, of determining how valid a conclusion is, as well as [understanding] its reasoning. Laboratory work is easier for me; it doesn't distress me. I have learned that if things

> can work out they *will* work out; if they don't work out it's not because of me. Before, I would get distressed because I felt that I couldn't do [an experiment] and that it should come out. Even though you still feel unhappy if it doesn't come out, you analyze the problem; you don't think you're stupid. (III/8th)

By the end of the undergraduate program, the students had internalized the scientific ideology and had identified with important aspects of it; moreover, they had developed a sense of belonging to a scientific group. Nevertheless, they were not—nor did they consider themselves to be—trained scientists. They would become scientists through their graduate studies and through work on different research projects. The thesis represented the first trial by fire in research for the students (and in teaching for the teacher/advisers). The following comments made by students indicate how they had internalized the undergraduate program's ideology:

> I feel that I'm going to make it. If they give me a problem, I feel that I am capable of solving it—of taking a project and running with it. What I don't feel capable [of doing] is looking for the problem myself. I feel that you can't be content with the idea that research is done, without having to reach the goal, because it's important for me to feel that I'm going to do something very valuable. . . . I get up feeling that I'm going to do something important, and I think that's important for doing research; I have energy. I also have little tricks to stimulate my creativity: I sit down to think of the title of the problem, and I leave it to my intuition; I think of fifteen problems and write them all down, and it works. The first thirteen are superobvious, the fourteenth is less so. Throughout the entire problem I look for analogies, and interesting ideas emerge. (III/8th)
>
> I feel better now because I am working on a specific problem, and I can gradually master it; I am able to discuss it, to propose things; I understand it more. Now [that I'm working on my thesis] I feel good, but there's more responsibility and you put pressure on yourself. (I/8th)
>
> I feel changes in myself. Before, I felt very insecure and it was hard for me to reason and to think fast. Now it's not difficult for me, and I come up with ideas; maybe I've learned to think fast. I

Table 11. General evolution of the undergraduate program in basic biomedical research, 1974–1990

	1974	1975	1976	1977	1978	1979	1980	1981	1982	1983	1984	1985	1986	1987	1988	1989	1990	Total
Entered	4	4	6	4	7	8	6	9	9	6	12	5	9	11	15	8	7	130
Finished coursework	4/100%	4/100%	6/100%	3/75%	7/100%	8/100%	6/100%	4/44%	8/88%	6/100%	9/75%	4/80%	3/33%	6/54%	...	...	...	...
Graduated	4/100%	4/100%	5/85%	2/50%	7/100%	8/100%	6/100%	4/44%	7/77%	4/66%	6/50%	4/80%	3/33%	...	...	...	...	...

	1979	1980	1981	1982	1983	1984	1985	1986	1987	1988	1989	1990*	Total
Graduates per year	2	8	3	...	10	7	4	7	7	3	8	5	64

SOURCE: Data were provided by the Academic Unit of Professional Cycles and Graduate Studies of the Colegio de Ciencias y Humanidades, Secretaría de Asuntos Escolares.

NOTE: The study population from 1974 to 1987 comprised a total of 100 students, of whom 79 (79%) finished all coursework, 64 (64%) graduated, and 20 (20%) did not complete their coursework; 1 student is now finishing the coursework.

* As of November 9, 1990.

think of alternatives; I have more imagination. I think that what is happening is that, before, I didn't get around in my field; now I know a lot and, therefore, I can think more. (I/8th)

By 1982, 8 out of 10 students had graduated from the undergraduate program and were pursuing graduate studies. By 1990, 64 students out of 130 who had entered had graduated from the undergraduate program. As shown in Table 11, a high proportion of each cohort's students finish their studies, a mark of the program's effectiveness.

6
The Acquisition of a Scientific Identity

Throughout this book we have described a process of socialization by which an individual is integrated into a social group. At birth, each individual must be socialized. This includes receiving from the surrounding group a system of symbols and meanings (culture) that explains the reality which the individual must internalize in order to understand the system, thereby giving meaning to reality and to him- or herself. This entails 1) a process of interaction with a group and with its role models in which important affective bonds are established and 2) a language and an ideology that will give a name and meaning to whatever is transmitted (Clausen 1968; Bock 1969; Elkin 1960; Aberle 1961).

Socialization is a dynamic process that continues throughout an individual's life. It is dynamic because even though the individual must internalize the culture that is transmitted through its socializing agents, it also modifies and reinterprets it (Giddens 1973). In this manner, society reproduces itself and continues and, at the same time, changes. Hence, the purpose is to internalize or appropriate the symbolic and affective elements that explain and reinterpret both reality and the individual within it, since in forming part of the group the individual adopts aspects that link him or her to the group and that begin to govern him or her from within the group.

We can distinguish two types of socialization: primary and secondary. Through primary socialization, the child is assimilated to a society and culture. This process is carried out, principally, by the child's family and teachers, who are the significant role models throughout childhood. In this process, language, behavioral patterns, norms, and values are internalized; as a result, one acquires an identity within the social group to which one belongs. Secondary socialization refers to the ulterior processes through which a socialized individual passes in order to form part of various specialized subgroups in society, such as the apprenticeship of trades or the acquisition of new functions. In secondary socialization the specific knowledge of a new role is assimilated; this includes the acquisition of a special vocabulary, new rules of behavior, and a particular vision of the world and oneself (Berger and Luckman 1976, 130–38; Clausen 1968; Bock 1969; Elkin 1960; Aberle 1961).

Some occupational functions require a more intense socialization, for they demand that an individual be unusually involved and dedicated. The purpose here is not that the individual will acquire yet another secondary identity, but rather that this identity will permeate his or her individual identity and life-style. Well-known examples of this type of socialization are the priesthood and the military career. Another example, presented in this book, is the role of the scientist. The scientist's socialization is similar to that of the above-mentioned professions with respect to the intensity of the process and the required dedication and loyalty to a special community. Indeed, the process of scientific socialization takes place via the individual's integration into one or several "families" of researchers: cohesive groups that have their own ideology, operating rules, and relationships. The process of integration into a scientific family is so intense and tightly knit that it resembles the sort of integration that occurs between a child and his or her family and society. Socializing such individuals requires mechanisms that are more like those which arise in primary socialization processes, in which the transmission of norms and values as well as a worldview has central importance.

Each social group that requires socialization such as the one just described utilizes ad hoc mechanisms, derived from the activity that is being conducted: members of the military, priests, and scientists all stress discipline, albeit with different meanings and modes of transmission, in accordance with the type of discipline required (physical, moral, and mental, respectively). In the case of scientists, socialization must be intense, since the purpose is to carry out an extremely difficult activity (viz., thinking systematically and unconventionally) within a strict but implicit system of control where recognition is granted only over the long term, even though recognition is precisely the chief motivation for scientists, who do not enjoy the incentives common in other fields (e.g., economic and hierarchic incentives). Moreover, this is a group that has extraordinarily tight bonds—often like family bonds—in which the predominant stimuli come from the group itself. For this reason, cohesion is essential, although it is maintained through implicit, often hidden, mechanisms. In addition, the high cost of training a scientist—in terms of the time and effort invested by professors and researchers—makes it important to prevent any candidate from dropping out and to ensure his or her productive integration within the scientific group.

Developing a Scientific Identity

Socialization is a complex process; it involves not only integrating new members into a social group, but making this group redefine itself and change in a certain way. Thus, there is a mutual influence between those who socialize and those who are socialized. In the case we have analyzed, there was a double socialization process, since the objective was not solely that students become scientists but that, through this process, professors might question and redefine themselves as researchers and teachers. Moreover, the development of three identities converged in this case: that of the scientist, that of the nontraditional professor, and that of an undergraduate program for training researchers which was new in Mexico and sui generis.

The undergraduate program's mystique emerged from the questions posed by the program's founders regarding the traditional method of training researchers. Gradually they defined the new scientist they wanted to form. Until this point, scientists had not criticized each other so much as they had criticized the educational system encountered during their own training. By coming into contact with students and by becoming role models for the identification and transmission of an ideology, the professors gradually began to question their own role as researchers vis-à-vis the need to present themselves as "good scientists and advisers" before the questioning eyes of their students.

As the students gradually internalized the scientific ideal and acted out some of its features (e.g., a critical attitude), they also began to question and to demand more from their professors, who in turn strove to act as strong, confident models, enamored of their scientific work; they reflected on their own behavior and ideal model. Through this, they began to modify their behavior so that it would fit the scientific image they themselves had created.

This then, was a socialization process in which students, gradually shaping their identity as scientists through the introjection of ideological, cognitive, normative, and affective elements they perceived in their professors, reflected back to them an image that, based on the ideal model, they had introjected. The professors likewise acted as a mirror (Lacan 1976) that reflected back to the students a positive or negative image, an image of recognition or disconfirmation as scientists (Laing 1969), as a result of which the professors themselves changed. This process was emotionally charged; the affective links between teacher/advisers and students assumed an enormous importance as an element of bonding, identification, and work.

The scientific image that was transmitted by the professors as a model of socialization was doubly idealized, for the professors wanted to make a reality of their own ideal image of the "good scientist" and, in addition, transmit it as a model to the students. This idealization entails a difficult process, filled with demands for both groups. It established very high standards to which the individuals aspired and expectations with which they compared themselves.

We understand "scientific identity" as the ideational and affective representation that one has of her- or himself as an individual who is devoted to scientific research and who is part of the scientific community. Perceiving oneself as a scientist—feeling that one is a scientist—that is, considering oneself to be equal to "them" is determined by the recognition one receives from the scientific community, which confirms that one is equal (Laing 1969) and which is based, in turn, on sharing an ideology, attitudes, behavioral norms, as well as a way of working.

Identity is a process through which the individual "aspires to shape his/her own ego analogically as a model" (Freud 1921, 2585). It is developed through partial identifications with model traits (Freud 1921), and it may take place at a conscious level (trying actively to imitate traits that appear related or attractive, trying not to assume traits that are valued negatively) or at an unconscious level (absorbing ideal models, seeing within oneself traits in common with the group in question, defining oneself in opposition to negative reference models, etc.). Thus, this is a process in which the individual assumes for her- or himself an aspect of the other, a process which transforms one, totally or partially, vis-à-vis an ideal model (Laplanche and Pontalis 1974, 191).

The individual develops through a series of identifications with others. It is through a long play of specular identifications that one gradually develops a concept of oneself, a feeling of personal sameness and difference. Through the identification process, the individual slowly shapes an ego ideal, an ideal to which one always aspires and which governs one's conduct and expectations; this ego ideal is made up of identifications with cultural and parental ideals as well as with significant others.

Hence, we can distinguish two levels in the identification process: a first level, at which the individual incorporates aspects of the other into her or his ego, which she or he adopts and makes her or his own; and a second level, at which she or he substitutes or completes her or his ego ideal through the ideal of the other. These two aspects are described by Freud when he analyzes the psychology of the masses, distinguishing between the identification that takes place between members of the masses and that which takes place with the figure of the leader (Freud 1921, 2603). When analyzing the process of identity formation, it is important to take these two aspects of identification into account in order to become familiar with those aspects which make the individual perceive her- or himself as equal to others as well as those aspects which govern her or his aspirations to emulate others.

An individual's identity is developed through his or her real and imaginary relationship with others. It includes aspects of one's uniqueness as well as aspects of difference, aspects that are molded through the relationship of empathy (Kohut 1980) and union with others, and it is based on this relationship as one's difference is gradually established.

Although we speak of "identity," we should pluralize this word. For an individual is not removed from his or her social surroundings; rather, it is in relationship to these surroundings that an individual gives meaning to him- or herself and to others, in a time and space that are frequently changing.

Hence this feeling and concept of sameness is joined by partial identifications with the traits of individuals and with social and ideological groups. The individual has various identities at any given time, with different levels of consciousness. Throughout one's interaction with others the individual gradually feels similar to some, different from others, and perceives him- or herself in terms of an internalized language, based on the similarities and differences that are gradually perceived vis-à-vis others. It is through this process that models are transmitted and affirmed.

The acquisition of a scientific identity is the result of a long, dynamic process in which the interaction among professors and students, as well as the acting out of the different facets of the researcher's role, allows students to assimilate the set of behaviors that a researcher has, as well as her or his ideology. Young reseachers gradually begin to feel and perceive similarities between their own ideas, emotions, and behavioral traits and those of their professors, and they compare the differences (Grinberg and Grinberg 1980) through identification with scientists, which constitutes one element in the development of an identity.

Identification does not just take place via behavioral traits observed or expressed; it also, and especially, takes place via the ideals of significant models, through the messages professors send regarding what they believe and what they would like to do. Thus, Freud (1914) pointed out that an essential formative factor in a child's identity is the identification of the child's superego with the superego of his or her parents, and not necessarily with their actions per se. Freud (1921), moreover, describes as a fundamental aspect in the constitution of the group the identification that takes place in each individual through internalizing and appropriating the leader's ideal. In the case we studied, students gradually made their own ideals of their teacher/advisers and governed themselves through them. In this study, we noted that a fundamental aspect of the process of scientific training was precisely the integration of the scientific ideal that the professors had internalized into the students' ego ideal. Students identified not only with the real and ideal image that they had of their professors, but also with the image that their professors reflected back of how they perceived the students as quality reseachers, and they confirmed this identity as being true (Bleichmar 1976).

All this eventually leads to an individual identity—a representation of oneself as a scientist—and to a social and cognitive identity (Merton 1976), through which the individual feels integrated into the scientific community and an area of science or a theoretical current, while at the same time she or he is recognized by the community as one of its members. In the case we studied, students entered the undergraduate program with an idealized image of the scientist, whom they saw as having quasi-heroic qualities. (Upon entering the program, one student told us: "I am convinced that scientists will solve our country's problems.") This idealized image was complemented by the transmitted image of the undergraduate program as a very high-level, exclusive group. It was this idealization that allowed the students to endure the rigorous initiation process, since the high valorization of the program and of the professors was also linked to the high self-image of the students who had been accepted for training. Thus, to the idealization of the program was added an overvaluation of the student.

The student relationship with teacher/advisers was important not only for the students but also for the professors, who experienced the possibility of fulfilling their ideals through young people—their "children." The professors projected to the students their idealized image of scientists and stressed the traits of this image through their evaluation of student behavior and in their daily interaction. Through such professorial projection, both overt and concealed, the students began to identify with scientific qualities in a positive way (trying to acquire these traits) or in a negative way (rejecting them or identifying with the opposite traits). Students imitated the behavioral traits that they observed and saw reinforced by the milieu. Moreover, the activity of research per se gradually demanded and molded the forms of behavior that the students developed by following the image of their professors or by being supervised by them. In this way, two fundamental aspects of the students' identity were formed: 1) a manner of being and acting as scientists and 2) an ideal that governed and controlled the students, to which they tended to aspire (this is what Freud [1921] distinguished as ego and ego ideal).

In the case analyzed here, the socialization process and, with it, the development of an identity arose initially through the emphasis in separating scientists from aspirants. Although the manifest language stated that students and professors formed a community and that students were considered to be—and worked from the beginning as—scientists, the reality was different. At the beginning (from the introductory course through the second year), professors set up a barrier, preventing dependent relationships and close ties from emerging between themselves and the students. In contrast to artisanal trades and other professions in which the student/apprentice forms a close relationship with the teacher in order to learn from him or her, in the case of scientists this learning is carried out in a second stage. The beginning is rigorous, hard-fought, and lonely; through it the students must show themselves capable of winning a good relationship with a teacher/adviser. The implicit promise is the possibility of coming to belong to a scientific "family" via acceptance by the "parent/adviser."

In their initial relationship with the students, professors also emphasized difference: in addition to maintaining a social distance which prevented rapprochements that could have led to dependence, they showed the students what they *were not* and what they *did not have* (e.g., they were not sufficiently critical, aggressive, inquisitive). Moreover, this idealized image of the scientist and of the group to which the students hoped to belong contrasted with the image the students gradually formed of themselves as scientists, vis-à-vis the difficult situation they faced in this initial stage, since they were experiencing the difficulties and frustrations of experimental work and listening to their professors' criticism of their mistakes. This image they had of themselves was counterposed to the initial overvaluation: the students felt they were a select group because they had been accepted into the program; it was through the emphasis on what they didn't know or were lacking that the professors conveyed their ideal of the scientist and the students gradually identified the personality traits, values, and modes of behavior prevalent in the group they wished to join. And it was at this point that the game of similarities and differences between the two groups began. This discourse was uttered by a scientific group known for its excellence and legitimated by the community, which marked the boundaries between the two groups—scientists and aspirants—even more dramatically and buttressed the idealized image of scientists and the difficulty of becoming one, as well as their desire to do so.

The above can be contrasted with the overt message of equality, which treated aspirants as researchers. The two messages were integrated, as the possibility of becoming a recognized and accepted member of the group was stressed, and as the quality and importance of the group to which they might belong was stressed. Thus the professors would comment that the Institute was a center of excellence and quality; the program's academic and teaching characteristics would reinforce this concept.

Moreover, an idea was circulated in discourse concerning the scientist's special gift or vocation, which cannot be described but which is nonetheless developed. This idea, together with the desire of being identified as a member of the group and part of a family, as well as the pleasure derived from the activity of doing research per se, helped the students to tolerate the frustrations they experienced, especially in the first stage of training. Later on, other factors came into play, such as the desire to learn and relationships with the scientific group or family, which mitigated the feelings of frustration.

When they perceived that the students were showing a desire to belong to the group and that they were assimilating to a certain extent scientific values and norms of behavior, the professors allowed a new, closer sort of relationship, in which an overt dependence of the student on his or her teacher/adviser was permitted. In this period, students began to imitate the behavior of scientists and to include in their own discourse elements of the scientific ideology; hence, it was possible to perceive an internalization of the scientists' ideal, as well as their training and behavior. The professors responded with an attitude of acceptance, although they continued to point out the students' shortcomings. More than pointing out what the students were not, they would comment on where the students were lacking: "They still can't do an independent project"; "they swallow up everything they read or hear."

The game of similarities and differences in interaction throughout the training process is complex. The very adhesion to scientific ideology, in addition to favoring similarity to the community through internalization of the ideal model, stimulates the acting out of the difference, by highlighting the importance of individuality and originality.

Stages in the Process of Identification

As soon as they entered the undergraduate program, students joined a group which changed in closeness and inclusion; they joined the Institute's scientific group as the socialization process advanced. Initially, the students were integrated into a cohort that belonged to the Institute, which was considered a select and privileged group within UNAM, and one to which numerous high-level researchers would be devoted.

During the third and fourth years of the undergraduate program, students joined a "family"; that is to say, a "laboratory" made up of a closely knit group of researchers. At this stage the students began to be integrated into the scientific group per se, until the point where they belonged more to it than to any other group, thereby breaking their ties with their cohort companions. Identification occurred with the group and with the teacher/advisers who led it. Students learned the rules of the game for the laboratory and the social norms of the community in general, all of which were permeated by the ideology. In the laboratory group they concentrated on playing the researcher role as well as on learning how things are done and achieved.

Beginning in the second half of the fourth year and continuing through graduate studies, the students joined networks. At this point, they would begin to assume a place in the broader scientific community, through their own and their teacher/advisers' performance in scientific work. In this manner they made contacts through publication, scientific congresses, and similar forums.

Moreover, we can distinguish two stages in the transmission of scientific ideology:

1. Transmission of the scientific ideal (first and second years)
2. Molding (third and fourth years)

Although these two stages coexisted from the beginning and throughout the scientific training process, we were able to observe a change in emphasis. Thus, although at the beginning there was a molding of forms of behavior, the transmission and internalization of the ideal predominated during the first two years. Professors consciously and unconsciously transmitted an ideology that included their personal ideal and the scientific ideal. Students internalized the ideals that were transmitted and the common traits that gradually stood out, as well as the negative points they detected which they did *not* want to acquire.

At the end of the first stage, students had identified with the researcher's role: they considered themselves researchers and behaved accordingly, although they made numerous mistakes. As we noted in the preceding chapter, the last two years of the undergraduate program reflected an internalization of the ideal; the predominant concern was with molding and affirming behavior. In this second stage, the emphasis was on fine-tuning what had been internalized, through the feedback students received from their own work as well as through their interaction with different groups.

In addition, we can distinguish five stages the students went through during the socialization process:

1. Powerlessness or a decline in self-esteem (introductory course and first year)
2. Overvaluation (second year)
3. Questioning (third and fourth years)
4. Integration (starting in the fourth year and, especially, during the thesis work and master's program)
5. Consolidation (doctoral and subsequent work)

In each phase small crises would arise.

The process of socializing scientists implies an apprenticeship period, as described by Berger and Luckman when they write about internal processes of secondary socialization and about what they call the "processes of alternation or resocialization," the purpose of which is to transform the individual. For this to occur, the individual must remain in the socializing milieu at all times and must be isolated physically and mentally from the previous milieu (Berger and Luckman 1976, 158–59). In the case of scientists, although the socialization process is less drastic, it has traits similar to those we described above and explains the emphasis on ideology.

First Phase: Powerlessness or a Decline in Self-Esteem

During the initiation period, or "apprenticeship," the demands were so high that they forced the students to set aside other interests and hobbies in order to devote all their time to the undergraduate program. In addition, and owing to this, there were very few contacts with the "outside"; students stayed at the Institute all day. During this stage, one of the few outside visits the students made was to an occasional scientific congress, the principal purpose of which was to broaden their range of models and reinforce their sense of belonging to the community.

The professors would emphasize the fact that the students were in a trial period, and they would frequently convey the message that this was just one difficult stage among many others the students would have to go through in order to become good scientists. In this first stage, then, students had to prove they had the potential for being researchers; for this reason, they were quite sensitive to indications of rejection by their professors, who were waiting for the results of this trial to begin to establish a close relationship, one of recognition.

Although the students had begun to feel that they were a select group because they had been admitted into such a special program, they now began to doubt their abilities and to recognize their shortcomings. Similarly, as they gradually discovered and assimilated the elements that, in a rudimentary fashion, make up the scientific ideal, they felt ill-equipped and doubtful regarding their potential to become researchers. They visualized the research degree program as full of obstacles and much more difficult than other degrees.

The students who did not drop out at this stage began to value the—at times painful—effort they were making, and they esteemed the research degree above all other degrees, as they felt that the training was more valuable and of a higher quality. This idea helped them remain steadfast in their most discouraging moments, which were frequent at this stage; in addition, their desire to learn and their curiosity (which is extraordinary among the students in this degree program) helped them find encouragement in the experiments and in the discussions they had with their professors.

The students had to show their professors that their abilities corresponded to the model of the ideal scientist (see Chapter 4, above), with which they were relatively unfamiliar; they would select elements that they were beginning to grasp from the ideology being transmitted to them, and they would try to act these elements out (generally in an exaggerated manner). The professors' response was not one of encouragement or recognition, but one of criticizing and demanding without providing clear guidelines on the models to be followed or the behavior that was expected, and this caused confusion among the students. At the same time, the professors maintained a distant, judgmental attitude, although they were in fact very involved in the students' development.

During this first stage, students felt they were being underestimated, and they were insecure; they strove to work better and show what they could do; they were frustrated at not meeting the demands and at feeling they were not "making the grade" vis-à-vis the image they had formed of the ideal scientist. For these reasons, some students dropped out.

Moreover, the numerous demands overwhelmed the students, who felt drained and unable to do what was being asked of them. They perceived the scientific world as being so demanding and fierce that it would devour them. Thus, they began to fear that they would be absorbed by the group and lose their individual identity; they were afraid of "becoming alienated" in the Institute from having to meet so many demands and remain there all day. Zinberg (1974, 242–53) observed something similar among English first-year chemistry students. This fear disappears during the following stage, when the students begin to identify with their professors. Eventually, by the end of their university studies, students come to assume the forms of behavior they originally feared and disapproved of—those of "alienated" scientists—without showing any displeasure or criticism.

At the end of the first year there was a change of attitude among the professors, who for the first time conveyed to students their relative satisfaction ("things aren't going that badly"). This helped the students tolerate their frustration and encouraged them to move ahead, in addition to instilling a liking for research, which was an important motivational factor.

Second Phase: Overvaluation

A close relationship with the teacher/adviser arose in the second year, as a result of the students' efforts to adopt the attitudes that had been stressed so much and because of the professors' recognition of the students' abilities. In this phase, students began to play the role of researcher, and the professors treated them differently, addressing them as "young researchers." The emotional climate contrasted noticeably with the discouragement and frustration of the first stage. A certain confidence began to arise among the students, as well as a satisfaction from feeling they were acting as researchers. Initially, the role-playing was apparent: students began to "propose experiments" (even though many of the experiments had been suggested by the professors during discussions). Slowly, their acting out the role of researcher became more real and complete, and they even proposed problems and designed experiments. The professors' supervision was indirect, which made the students feel that they were working with their own ideas and contributing something to their professor's research group. During this stage, professors presented a different attitude toward the students than they had previously. In recognizing them as potential researchers, the professors gave them their support and made them feel that they were part of the research group, inviting them to participate in group seminars and, at times, to work in the same laboratory. Students began to consider themselves researchers. Professors gave the students positive feedback and openly supported them.

At the same time, the professors established more-open relationships with their students: they allowed the students to see them at work, and they even talked to them about their private life, although they always protected their image when doing so. The professors came across as strong figures who were sure of and in love with their career, and who were very dedicated to it.

The students imitated their professors' behavior and began to consider themselves researchers, which gave them security and satisfaction. By this time, they had partly internalized the ideal model: they thought of the scientific career as heroic; they idealized their professors and the work itself. The students had assimilated elements of the scientific ideal, which emphasizes original, creative thinking, questioning, and inquisitiveness. All these factors stressed the development of individual abilities and concentration on one's own work. This favored an immersion of each student in her- or himself which at this level was taken to an extreme: students tended to scorn whatever came from outside—even, for example, scientific tradition. Thus they felt it was more important to look for a creative idea than to master a topic. This attitude was encouraged by some professors.

At this stage we could detect a certain delight among the students vis-à-vis their own image and behavior, which, as just noted, were centered more in themselves than in others. Even in the discussion seminars, the students talked of there being a "struggle of personalities"—more than an exchange of ideas and information. This delight was related not only to their own image, but also to their image of science. The students proposed grandiose experiments, saw the scope of science as being all-inclusive, and felt that they were consummate researchers working in one of the country's best research centers. The professors concluded the overvaluation stage by making the students confront their inexperience as researchers, their deficiencies, and their shortcomings.

Third Phase: Questioning

When the professors forced the students to examine their shortcomings, when they criticized the students' work and the exaggerated way in which they had begun to follow what they perceived as the ideal model, the students again felt that they were being underestimated; they also became insecure, for they realized there were many things they had not yet mastered. Moreover, the students were disappointed when they realized that the road to becoming a scientist is a very long one and that they had a long way to go before they would become scientists. They also found research to be somewhat routine and monotonous, and they became bored. Too, they began to notice shortcomings in their idealized models when they observed their professors, who did not always measure up to the ideal model.

Gradually, though, the students began to accept their own shortcomings and difficulties and, in so doing, perceived themselves as potential researchers in spite of their deficiencies. The students developed and integrated the ideal, even with the faults they had found in their professors and in themselves.

Fourth Phase: Integration

The development and integration of the ideal, which began at the end of the fourth year, continued while the students were doing their thesis work. During this period, students began to act fully like researchers. They gradually integrated their own model as researchers, which was structured through partial identification with the different traits of their professors or with those which they had experienced, assumed, or imagined.

The students began to accept scientists in a less idealized form, acknowledging their shortcomings: "Scientists aren't so perfect"; "research isn't so easy." Their criticism of professors and of scientists in general diminished, because "everyone makes mistakes." Students worked on their own problems.

As the end of the fourth year approached, the students were pressured to decide on a research project and a teacher/adviser and, likewise, to make a greater commitment and identify more with a given research group or field. This decision was crucial, since it implied choosing a line of work, which would have repercussions on the students' academic life; the relationship with the teacher/adviser was, generally, very close, and the area the students chose became their field of specialization. They would thus acquire a new and distinct identity as scientists.

At the end of this phase the students began to integrate the ideal model, which was established as an "ideal of the ego" of what they should become. The shortcomings they had found in the professors' and their own behavior, in addition to the problems of everyday research life, were set aside.

Fifth Phase: Consolidation

The socialization of the scientific career was not concluded in the undergraduate program. Indeed, this is a neverending process—one that evolves throughout each researcher's professional development. Nevertheless, we feel that when the student completes his or her doctorate he or she reaches a point of relaxation: all the ideological elements have been integrated into an ideal, and some have been applied to the daily task of scientific work.

Discussion

Throughout the socialization process each student developed his or her own image of a scientific identity, based on elements the student experienced and received from professors and colleagues, and from publications and scientists in general. Although the students spoke about the traits they wished or did not wish to acquire as scientists, they were unaware of many traits that they had already internalized and that guided their behavior. When we questioned them on this point, they responded favorably. Hence, we must admit that in the cases studied herein, conscious and unconscious elements played a role in the formation of the students' identity. Therefore, we disagree with Bucher and Stelling (1977, 175–76), who conceive of professional identity as an active construction in which students consciously select traits in order to develop their own ideal models.

Throughout the socialization process, individuals are trained and learn to work in the production of knowledge, to be members of a research group; they internalize the community's ideology, which allows them to interact and communicate with other scientists. It is interesting to point out that in those cases where there was a prior socialization, the process went faster. Hence, those students who tended to identify with the idea of the scientist that the program was trying to form became more socialized, and those students who did not completely agree with the image had greater difficulty integrating themselves into the group and into the degree program in general. Thus the process was, to a large degree, self-selecting.

The scientific identity develops gradually, through the resolution of different crises that generally stem from the confrontation of what has been experienced with a conflicting ideal that has been formed. The manner in which the students gradually resolve such conflicts produces a response of recognition or self-criticism from the group and from the teacher/advisers, which eventually confirms or disconfirms the students' image of themselves as scientists. A similar development occurs in the case of the socialization of sociologists, as was reported by Reinhartz (1979), who pointed out that the manner in which the student gradually resolves the differences she or he finds between the ideal of what it means to do research and what she or he has observed or experienced characterizes the student's identity.

Thesis work is an important stage for managing the conflict between the ideal and what actually occurs, just as role-playing, with its different levels of involvement according to the moment in the process, was an important stage. Studies have highlighted the importance of role-playing for the development of the identity of teachers (Pavalko and Holley 1974), doctors, and biochemists (Bucher and Stelling, 1977).

The preponderance of the teacher's role in the socialization of students has been a controversial topic. Hagstrom (1975) and Merton et al. (1973) pointed to it in the training of scientists and doctors, emphasizing the impact professors have on students' attitudes and on their choice of a field. Reid (1981) felt that teachers were the greatest influence in the case of the socialization of dentists, and he found influence in the adoption of teachers' ideals, based on the perception students had of their teachers as persons who supported and wanted to help them.

The evaluative attitude that Bucher and Stelling found in their study (1977) was also seen in ours, and it can be understood within a context of interaction with professors, who evaluated themselves and who evaluated their students and colleagues as well. As we have seen, such an attitude is a source of ideological communication and identity formation, for it is an element that creates feedback. We acknowledge that identity is formed only through others, that an individual identifies with the real or ideal image that she or he perceives in others, and that the individual depends on these other persons to maintain the identity as she or he responds affirmatively to changes in the development of her or his own identity (cf. Bleichmar 1976, 51–54).

Bucher and Stelling give little importance to teachers, who they felt fulfill only one support function in the training of biochemists. Furthermore, they determined that the identification with the field and the experience of mastering the area of specialization are essential (Bucher and Stelling 1977, 185). For these authors the most important variable is role-playing, since, in acting like researchers, students felt they had mastered the discipline, and this gave them a feeling of achievement. In addition, they felt that the experience of mastering the field confirms a student's identity, as students do not need to feel validated by their professors. They found, however, that students constantly evaluated themselves and also evaluated their classmates and professors (Bucher and Stelling 1977, 175–268). As can be seen in our study, role-playing is not sufficient for acquiring a scientific identity: a close relationship with advisers is also essential. The lack of coordination of these two variables had consequences, in the form of setbacks or the slower development of an identity.

Conclusions

The socialization of scientists in peripheral countries occurs in "adverse conditions" compared to those that exist in advanced Western countries, where science developed naturally and in response to those societies' internal needs (Bernal 1979). Developing countries, generally speaking, lack scientific traditions and, therefore, the scientific ideology is absent from the primary and secondary educational systems. Hence, basic scientific attitudes and values—necessary for carrying out research activities—must be acquired by individuals late in their academic careers. These "adverse conditions" are the result of various factors, first of which is that the scientific community developed late (see Chapter 1 on the development of science in Mexico) and in a situation of cultural dependence on foreign centers. That is to say, science did not result from internal development; it is an imported cultural product.

In the modernization process that developing countries go through, there is an awareness of the interrelationship between science, technology, and industrialization. For this reason, scientific research is perceived, albeit not clearly, as an integral part of development. Thus scientific research is accepted and imported, along with other elements of the same process, even when local educational and research institutions are not prepared or suitable for bringing about the support and stimuli needed to carry out this activity (see Ben-David 1968 on the effect institutions have on the development of science).

In Latin America, scientific research faces a shortage of resources and a lack of established mechanisms for recognizing scientific achievements, all within a sociocultural atmosphere that suffers from an insufficient development of the scientific tradition. This is reflected, for example, in the fact that the central objective of higher education is to teach students to use, rather than produce, knowledge. Consequently, research continues to be isolated from production and, in general, from cultural development.

In the case studied here, a program was created for training scientists in a "university for the masses" that is devoted to training professionals. The students are the product of an authoritarian educational system in which one is taught to accept truths more than to question them (Paradise 1978; García and Vanella 1992). The group of researchers who created the undergraduate program saw it as a special program, one in which—in contrast to what usually occurs at UNAM—groups would be small and rigorously selected; they were imbued with a profound sense of mystique and a desire to begin forming Mexican scientists who would be highly motivated and well trained from an early age. They saw the program as a break from the existing academic tradition and, therefore, as something revolutionary within the national educational system, since, for the first time at UNAM, the social production of scientists would be headed by researchers themselves from the outset of the undergraduate program.

All this lent the early years of the program a special mystique in which the transmission of a scientific ideology or ethos had great importance. Indeed, at the outset of our case study we did not intend to stress the ideological aspects of training scientists; this gradually emerged through the field material, from the insistent discourse of the professors and, eventually, from the students themselves.

Many questions have been raised about whether the universal scientific values Merton speaks of (1973a) actually exist or, rather, whether they predominate because they are concentrated in certain scientific training centers (in developed countries) which therefore share an ideology. Our own analysis indicates that the scientist's very manner of working requires particular characteristics and ways of being which also influence the assumption of certain values. Science's work-style and objectives translate into universal values that allow scientists to share ideas and communicate with each other.

Nevertheless, there are peculiarities to national scientific communities. Latin American scientists feel that, more than national characteristics, there are socioeconomic conditions that distinguish science in developed countries from science in the Third World (Leff 1979a and 1979b).

The program we studied attempted to train researchers in a way appropriate to the problems and needs of a Third World country where scientific work develops in difficult and at times adverse conditions. The scientists saw the program as being on the cutting edge, and they placed high expectations on it. The program itself was a challenge and an ideal for the professors. This meant that the scientific ideology, which integrates a highly idealized model, was combined with a second idealization regarding the meaning and the fate of the new program.

The high degree of idealization in the scientific mystique entails several problems. The ideal image of the scientist implies perfection; the over-idealization of this model, therefore, makes subjects aspire to an unattainable ideal.[1] The ideal model of the "good scientist" implies the existence of its antithesis, the "bad scientist." Thus, overidealization of this model makes the subject operate within a binary logic of all or

---

1. Dr. Bertha Blumm discussed these concepts with us, offering useful insights.

nothing, of the good scientist versus the bad scientist, in which one may feel either omnipotent or impotent, depending on whether one feels he or she is or is not approaching the positive ideal. Moreover, expectations can be created that are so high they are too difficult to attain, causing a genuine achievement not to appear as such, since achievement leads to evaluation vis-à-vis the ideal, which will always make what has been done appear insufficient or imperfect; thus, there is the risk that the individual will founder in the face of a mistake she or he experiences as a narcissistic wound (cf. Bleichmar 1976). This implies living in permanent tension, with the individual always aspiring to achieve but fearing to fall into the abyss of imperfection.

From this the need to mitigate overidealization stands out, as does the need to provide the possibility of small achievements that will confirm a subject's identification with the positive, rather than the negative, ideal. Thus, the mystique and charismatic personality of the teacher plays an essential role in training the scientist.

Another consequence of the specific institutional conditions of the "university of the masses" was that there arose a need for the program to become closed and turn into a kind of monastic sect in which, owing to a shared mystique, invisible walls were built. Coser (1978, 101–5) talks about certain "voracious institutions" that exist in modern societies, that demand the unwavering adherence of their members and attempt to envelop their personalities. These institutions depend on the voluntary adherence of their members; therefore, they develop mechanisms to trigger the members' loyalty, and they erect symbolic barriers between them and society, stressing the separation between "them" and "us." Sharing a strong ideology is part of their differentiation model. Although the program we studied did not go to the extremes described by Coser, it did share some characteristics: for example, pressuring students to weaken their links to external activities that might compete for their loyalty and transmitting a mystique that would give students an awareness of scientists being different from members of other groups— an awareness that ceased to be explicit after the early years, though never less intense. The need to train young scientists in an environment that is so closed to external influences is questionable. This problem has been gradually solved through the program's growth, which has begun to allow a larger number of students to be admitted, although the groups have remained small.

Nevertheless, this academic project, which began at the undergraduate level, was successful in forming what is now a large group of young scientists in the field of biology within the university. From 1980 to 1989 the program had 409 students (89 undergraduate, 212 master's, and 108 doctoral [Tapia 1989]). In this manner, the problem of social production in a research field at UNAM began to be overcome. The program we have described has succeeded in transmitting to the students an individual and social identity as scientists, even though it also produced some problems owing to its closed nature, which had been permitted in order to overcome the adverse institutional and cultural conditions in which the program had to develop.

Appendix 1

Curriculum of the Undergraduate Program in Basic Biomedical Research, First and Third Cohorts

The official program, presented to the University Council, is made up of the following areas and theoretical-practical units:

1. Area of basic knowledge: biochemistry, physicochemistry, biomathematics
2. Area of structure and functioning of living beings: histology, cellular biology, development biology, physiology 1 and 2, molecular biology, general biology, genetics, immunology
3. Area of basic methodology in biomedical research, I: biostatistics, research seminars 1 through 4
4. Area of general laboratory methodology: laboratory techniques 1 and 2
5. Area of pathological human biology: introductory course in medicine, general pathology, microbiology, virology, toxicology
6. Area of ecopathology in Mexico: special pathology 1 and 2
7. Area of basic methodology in biomedical research, II: research seminars 5 through 7
8. Area of special laboratory methodology: clinical-laboratory techniques 1 through 3

Although these are the officially recognized courses, the program was not taught in strict accordance with this description. In actual practice, some subjects have been canceled or complemented with others, as required by the needs of the teaching method. The objectives of the different areas, as indicated in the pamphlet, are as follows:

Basic Knowledge

- To use the concepts and methodology taught in the basic courses in experimental biology

- To use bibliographic information on the methods and techniques for conducting experiments that provide relevant information on a chemical or physicochemical phenomenon
- To select a methodology from among the various methods for the evolution of an optimal problem according to the following criteria: its precision; the quantity of information it provides; the cost, time, and possibilities of fulfillment from a theoretical and technological point of view.

Structure and Functioning of Living Beings

- To use bibliographic information on various methods that allow research to be conducted on aspects of human biology at a molecular, cellular, and organismal level
- To analyze the concepts and methodology of the study of normal physiological and biochemical processes through molecular and organismal approaches
- To propose experimental designs for solving biological problems
- To apply laboratory techniques in employing the proposed designs
- To analyze the experimental results in the context of current knowledge
- To organize the design and the experimental results for scientific communication
- To identify biomedical problems that arise in the development of an experiment and that can be researched technically or experimentally

Basic Methodology in Biomedical Research, I

- To identify problems of a biomedical nature
- To present these problems in operational terms so that they can be researched theoretically or experimentally
- To apply the heuristic method in the solution of biomedical problems

- To analyze the logical structure of experimental research
- To explain the process of conceptualization in the creation of scientific experiments in biology
- To evaluate the general structure of a scientific project
- To analyze the process of developing formal models in biomedical science

General Laboratory Methodology

- To explain the foundation and bases of the methods and instruments routinely used in scientific research
- To apply basic experimental work techniques efficiently regarding the cost of material, the optimal use of instruments, and time
- To select and summarize pertinent information in order to develop a short-term experimental plan

Pathological Human Biology

- To apply the principles of biological organization to biomedical problems, from the molecular level to the organismal level
- To explain the general pathological processes as forms of biological coexistence
- To use bibliographic information on a problem related to pathological human biology

Ecopathology in Mexico

- To analyze the most frequent ailments in Mexico from the standpoint of the sick individual and the ecological conditions that lead to these ailments
- To identify the socioeconomic consequences of these ailments

Basic Methodology in Biomedical Research, II

- To identify problems of a pathological nature
- To describe these problems in operational terms, in such a way that they can be researched at theoretical-experimental levels
- To analyze the basic methods of research on informative macromolecules
- To apply special methods and techniques pertaining to the study of population dynamics, from the molecular level to the social level

Basic Laboratory Methodology

- To develop and apply clinical laboratory techniques

Appendix 2

The Formal Knowledge That Was Taught

First Year

First Semester: First Cohort

The first semester covered the following subjects: physicochemistry and laboratory; biochemistry and laboratory; biomathematics; research seminar and general seminars. These courses included problems of biotechnology and public health.

In physicochemistry, closed, open, and isolated systems were studied, as were the laws of thermodynamics, the concept of entropy, thermodynamic potentials (Gibbs's free energy), and the equilibrium constant. Likewise, the applications of these concepts—e.g., diffusion, chemical and physical equilibrium—were covered. In biochemistry, atomic theory was analyzed: the structure of the atom and the levels of energy; electron distribution and the types of reaction mechanisms.

In mathematics, the concepts of event and the operational properties of independent events were studied; probability, probability models, combinations and permutations were covered, as were the different types of distribution. The geometric concepts of 1) derivative, integral, two-variable functions and 2) partial and total derivatives for two-variable functions were studied, in addition to exact differentials, maximums, minimums, and points of inflection for three dimensions. Minimum-square regression and principles of differential-equation solutions were also taught.

In the research seminar, the experimental method was analyzed through data analysis, questioning, and the search for alternatives. In conferences and seminars, some students were taught topics from the philosophy of science, economic theory, and sociology.

Techniques

The students became familiar with the laboratory equipment; they learned basic operations such as purification, analysis, and the identification and preparation of quantitatively and qualitatively controlled solutions. They used routine chemicals (general salts, acids, bases) and simple instruments such as test tubes, pipettes, and glasses of various shapes. They learned simple techniques and used some laboratory tools and apparatus, such as thermometers, scales, slide rules, temperature kilns, potentiometers, and general electronic intruments. Finally, the students were initiated in reading techniques, to enable them to study and manipulate information.

First Semester: Third Cohort

In the third cohort, it was decided that students would be introduced from the beginning to biological problems. The following seminars were taught: general biology, cellular biology, biostatistics, mathematics, research seminar 1, heuristics.

The general biology seminar was taught during the first half of the semester. Its objectives were to introduce the students to biology and to familiarize them with the laboratory and with experimentation dynamics. The cellular biology seminar took up the second part of the semester. The research seminar lasted the entire semester; it was designed to teach the students to analyze and criticize information. The mathematics and heuristics seminars also lasted the entire semester; the objective of the latter was to teach methods for analyzing and solving problems. Finally, four sessions of biostatistics were given in which a general overview of the discipline and its research applications were presented.

The general topics that were covered were the following:

- In biology: function and structure of the cell, mitosis, genic regulation, chromosomes, immunology, types of blood cells and their function
- In mathematics: analytical geometry, systems theory, formal logic

The students also designed experiments for studying an immunological problem.

Techniques

The techniques that were covered included washing and sterilizing glass material, blood smears, obtaining and distinguishing chromosomes in lymphocytosis, development of karyotype and sexual chromatics, observing mitosis, preparation of culture media, observation under a compound microscope, microphotography, computer programming, and graph interpretation.

Second Semester: First Cohort

The following subjects were taught in the second semester: cellular biology 1, research seminar 2, physicochemistry 2, and biostatistics. The two principal topics were cellular biology and molecular biology.

The course in cellular biology constituted the core of the semester. Its objective was the study of the cell at a morphological and functional level as well as its internal functioning. Some basic aspects of cellular genetics and genetic-information regulation were covered; structure and function at the cellular, subcellular, and molecular level were studied.

The essential objective of the course in physicochemistry was to present, in a general fashion, topics important to biomedical research (e.g., thermodynamics). As this course and the one in the previous semester were the only courses on the topic throughout the entire program, an attempt was made to transmit a comprehensive knowledge of the subject matter.

The objective of the course in biostatistics was to present concepts regarding precision, reliability, certainty, and error. The professor tried to impart conceptual foundations of statistics that would allow the students to expand their knowledge of the discipline and, in this manner, raise and solve problems in the future.

The students were expected to bring together all the concepts they had learned, to apply them, and to know how to handle them in solving problems and planning experiments.

Techniques

We can classify the techniques learned in this semester as follows:

1. *Reading techniques.* The professors wanted the students to learn to distinguish relevant information and to discard the work hypothesis, methodology, and results of the information analyzed; to analyze information meticulously and carefully, without skipping any points; to assimilate information free of interpretation and to summarize it and draw conclusions. They also hoped to teach the students to evaluate any given article, to question it, to criticize it, and not to take it as the absolute truth; to analyze the credibility of data based on logic, technique, and a comparison with reality; to consider other possible interpretations.
2. *Experimental techniques.* The professors' objective was to teach the students to look for the technique best suited to the problem at hand; not to skip steps during an experiment; to keep work instruments orderly, clean, and accessible; and to perform and repeat carefully each step of an experiment, taking into account external or laboratory factors that might influence the experiment.
3. *Biological techniques.* The students had to learn simple techniques, such as to use syringes, draw blood, and differentiate blood-cell types, as well as counting and identification techniques. They were required to work on a live organism (a rabbit) so that they would learn the technique for handling animals and obtaining samples of biological material (see Chapter 3, above).

They also learned to prepare and preserve solutions; to look through an electron microscope; to use a centrifuge; to control and manipulate the variables that directly influence the organism being studied (temperature, light, time, etc.); and to employ statistical techniques, such as ordering, organizing, and rank ordering data.

Second Semester: Third Cohort

The students were very enthused as they began the second semester, especially because they were studying biochemistry and physicochemistry, knowledge which they felt the need to acquire because of their experiences during the previous semester. The professors attempted to impart general theoretical foundations related to biology.

The following seminars were taught: research seminar 2, organic chemistry, physicochemistry, and biostatistics. In fact, this semester corresponded to the first semester of previous cohorts, albeit with some changes. The purpose of the research seminar was to stimulate the production and formalization of ideas in a research project. The seminars on organic chemistry, physicochemistry, and biostatistics attempted to present a general overview of the subject and to lay down the basic theoretical knowledge and its application to biology problems.

Second Year

Third Semester: First Cohort

The following subjects were taught: physiology, biochemistry, molecular biology, research seminar 3, and laboratory techniques 3. Concepts from genetic biochemistry, physiology, and molecular biology were taught, including RNA, synthesis regulation, the genetic code, transcription, hybridization, and metabolic paths. All the subjects were included in a broad research unit, which was divided into three topics: genetic biochemistry 1, macromolecular synthesis, and metabolic paths.

Techniques

The students learned to do in vitro experiments, to cultivate organisms in liquid media, to measure amino acids and in general to handle more-delicate chemical reactions, to separate DNA and RNA, to extract nucleic acids, to develop graphs, to make chromatographs, to use radioactive techniques, and to handle instruments that measure radioactivity. Some techniques were taught directly by the professors; others were learned by the students themselves or with the guidance of other members of the research group; for example, the doctoral students showed them how to use certain instruments.

In general, the students conducted more-complex experiments. They learned some "tricks" in applying the techniques that their professors and their research group had taught them, and they learned others on their own. They acquired greater agility in handling apparatus quickly, in making smears, in verifying substance quantities, and so forth.

Third Semester: Third Cohort

The third cohort's third semester was devoted completely to the molecular biology seminar. Students learned concepts from cellular physiology, biochemistry, and molecular biology, such as the interrelationships between a nitrogenized and a carbohydrate metabolism, protein structure, enzymatic kinetics, genome redundancy, the structure of the genome (DNA), the shifting of biological-system macromolecules, regulatory DNA and RNA signals, genetic regulation and genetic-information expression, steroid-hormone action mechanisms, the virus, and cancer.

Techniques

The students used various biochemical techniques and developed a mathematical model for the regulation of protein synthesis.

Fourth Semester: First Cohort

The areas of human and microbial genetics, medical microbiology, and bacterial physiology were covered. Among the topics discussed during the semester were basic principles of microbial genetics, such as genetic-exchange mechanisms (conjugation, transduction, repression); mechanisms of bacterial recombination; the characteristics of mutation and mutant forms; the characteristics of *Escherichia coli* and phages, and lysogeny.

DNA replication, synthesis, and protein regulation (concepts the students should have grasped the previous semester but had not mastered) were studied. The students covered some basic concepts related to human genetics, such as heredity mechanisms and types, gene characteristics, and gene interaction with the environment. They also studied the physiology of malnutrition and ailments that affect metabolic processes.

Techniques

The students chose and designed their research method vis-à-vis a problem their professor had raised. They had to design experiments to determine which of two strains was lysogenic, and they had to discover how to induce lysis and which agents were appropriate. They devised experiments to produce mutants. They developed a research project concerning the relationship between genetic factors and malnutrition. They also became familiar with selection criteria for the method and techniques to be used, such as feasibility, likelihood of obtaining material, and simplicity.

During this semester the students were able to apply known techniques to the study of new problems (e.g., isolating and cultivating a bacterium): how to make strains grow; how to look for a sensitive strain in order to compare it with others and determine which among several cultures is lysogenic; how to determine what chemical agents allow phages to be released; how to tell which paths can induce lysis. They also learned to distinguish and identify phages, to produce a mutant, and determine where mutation occurs. To produce the mutant they had to select a mutagenic medium; some students used ultraviolet light, and others used a chemical agent.

They also isolated *Escherichia coli,* using biochemical and chemical methods (medical microbiology techniques). They learned cultivation-enrichment techniques: how to make one type of cell grow more than others; how to plant cells and how to plant in a specific medium; how to sterilize, purify, count, and isolate cell colonies. In addition, they learned "genetic screening" techniques; statistical analysis techniques applied to the study of human heredity; urine and blood analysis; and techniques for diagnosing birth defects.

Fourth Semester: Third Cohort

In the fourth semester two basic seminars, each lasting a quarter, were taught. These seminars dealt with development biology and human physiology. It had been planned that a seminar in heuristics would be taught; in addition, students attended a seminar on population dynamics.

The students received information on human tissues and systems (nervous, endocrine, circulatory); on reproduction physiology; on the organization of chromatin in higher organisms; on the regulation of protein synthesis at the level of transduction; on the regulation of RNA synthesis in higher organisms; on the possible relationship of unusual nucleotides in differentiation; and on genome organization. They were also taught some concepts from demographics, economics, and anthropology.

Techniques

The students learned such techniques as paper electrophoresis, column chromatography, and simple and gradient centrifugation as well as exploratory techniques in humans: the electroencephalogram, the electrocardiogram, measurement of arterial pressure, audiometry, exploration of the back of the eye. They applied known techniques to the study of physiological problems. They learned to interpret graphs and had to plan the sequence of a research project. They improved their skills in developing protocols and focused on intermediate research steps.

Third Year

Fifth Semester: First Cohort

One of the main objectives during the third year was to expose the students to work-styles and research projects other than those seen at the Institute. The students were to have some contact with research applied to medical and biotechnological problems.

In this fifth semester two courses were taught: human medical pathology and an introduction to bioengineering. The students acquired specialized information on human medical pathology, a biomedical vocabulary, and a knowledge of different types of ailments. They examined some aspects of general pathology (e.g., cancer and pancreatitis), normal and pathological physiology of some organs (e.g., the heart, lungs, and liver), and surgical and histochemical pathology. They became familiar with some medical and biological problems in applied technology. They received a general introduction to the field of biochemical engineering. They also received general information on enzymes and learned concepts related to fermentative engineering. A basic overview was given of regulation mechanisms, microbiology, and metabolic generalities at the industrial, technical, and enzymatic level.

Techniques

The students developed a research project, from bibliographic review to experimental design. Their research was to cover the following steps: isolating a microorganism from the ground; studying its capacity, in order to see whether the proposed reaction was possible; finding the optimal growth conditions; and breaking the microorganism. They also learned to design fermenters.

The techniques they dealt with included the following: simple histological techniques, such as cutting and dying tissues; distinguishing in a microorganism a healthy tissue from one that has degenerated; carrying out autopsies; manipulating microorganisms in order to obtain substances having practical interest; FORTRAN programming; general strategies for metabolite enzyme hyperproduction through microorganisms; bioconventionals; microbial enzyme production; isolating and fixing enzymes; diverse forms of utilizing enzymatic activity; heat and mass transfer; and recovery of fermented products.

Sixth Semester: First Cohort

In the sixth semester the following subjects were taught: medical biology, immunology, human pathology, and parasitology. The objective of the immunology course was to expose students to the concepts and research methods in this field. The students had to work together on an existing project in their laboratory and discuss the characteristics and the problems of scientific work in Mexico. They dealt with concepts related to microbiology and applied them to the study of *salmonella* as an ailment and as a model for researching a bacteria-producing ailment. *Salmonella* was analyzed from a biological perspective (shape, method of culture, physiology, antigenic structure). In addition, elementary concepts in clinical immunology were covered, information was given on *cysticerci* and on different types of cancer as well as on onchocercosis, epidemiology, parasitology, bacteriology, and medical pathology, which were studied directly in the field by extracting the raw material from patients and using it in experiments.

The course on human pathology and parasitology lasted three weeks. It was taught at CIES (Centro de Investigaciones Ecológicas del Sureste), in Chiapas. The objective of the course was to expose the students to clinical research outside Mexico City, presenting them with different research fields and the problems implicit therein. The course was under the direction of three CIES professors, who had the students participate in the following activities: seminars on the research areas engaged in at CIES (immunology, entomology, bacteriology, and parasitology), for which purpose they had been sent bibliographic material beforehand; a seminar on socioeconomic problems (studies on migration among indigenous groups); and fieldwork.

Techniques

The students covered the following topics: techniques for isolating salmonella and other parasites; preparation of media for cultivating salmonella; immunology research techniques; methods for bringing about immunoelectrophoresis with serum, in order to detect whether a patient has cysticerci antibodies; reviving frozen cancer cells; inoculating animals; describing tumors; and measuring a patient's degree of illness or health. They studied the Bush techniques, and they learned to measure immunoglobulin levels and to extract dermatological samples from humans, to count microfilaria, to do electrophoretic profiles, to extract blood from humans, and to prepare an electrocardiogram.

The students applied various methods to develop or design a broad research project (how to mount an experiment on cancerous tissue, how to detect changes that arise in cancer patients). They had to develop a research project and lay out all the possibilities and alternatives (they had to choose the type of cancer they were going to experiment on, the subjects and the techniques). They also dealt with the administrative aspects of a project, such as obtaining and caring for research animals, obtaining supplies, staffing, and so on. At CIES, they faced the need to plan research using available resources.

Third Year: Third Cohort

In 1978 the undergraduate degree was divided into yearly programs based on research projects and complemented with theory seminars. The teaching commission felt that the time allotted to each course—from six to eight weeks—was not sufficient to conduct a research project, and it decided to organize the program into periods of one year.

A joint program was designed for the third and fourth cohorts. All students participated throughout the year in a molecular biology seminar. Moreover, each student had to choose a teacher/adviser and a research project in one of the following areas: molecular biology and development biology, which corresponded to three of the Institute's departments, with from one to three laboratories per department and their respective research groups. Nevertheless, the professors proposed to the students the possibility that they could participate in UNAM's Institute of Biology, which works in some of the areas mentioned above, so that they would become familiar with other types of research.

The students had two months to define their chosen area and link up with their teacher/adviser; to aid them in making their choice, students attended seminars with different researchers from the IIBM and the Institute of Biology who spoke of their own research. Later on, the students were to contact directly the researchers in whom they were interested; these researchers would suggest a bibliography and/or topics to work on. During this period, each student had to come to a decision regarding his or her project and teacher/adviser.

Fourth Year

First Cohort

The first part of the fourth year was devoted to an introductory course in medicine, which was given at the Hospital del Niño (affiliated with the DIF, Desarrollo Integral Familiar) so that the students could work on problems related to infection and apply their theoretical and methodological knowledge to those problems. The objectives were to make students feel that science should be at the service of society and to make them aware of the conditions in which research on humans is carried out, as well as its advantages and disadvantages as compared to laboratory research. The students acquired a basic knowledge of infection and an introduction to medicine. Some studied the digestive system, others the urinary or nervous system. In addition to the physiology, the symptomatic manifestations of infections in various systems as well as the antibiotics used to fight these infections were studied, along with virology.

In the second part of the year, it was determined that the students would focus on their research work (i.e., their thesis) and attend a seminar on the dynamics of human populations, the purpose of which was to broaden the students' knowledge in the field of general biology, introduce them to theoretical research topics in biology, show the impact of biology research on social problems, and allow them to consider the relationship of biology to the social sciences. Broad concepts related to physical anthropology, economics, demographics, and general biology were taught. In addition, the students learned concepts related to each of their research topics, as well as concepts related to genetic engineering, development biology, biomathematics, ecology, and virology. Moreover, a seminar was taught on selected topics from organic chemistry— at the request of the students, who felt behind in this field.

The students had to design a research project in a field related to infection and learn to diagnose using the largest amount of information possible, making use of diverse sources. They had to learn how to extract information from clinical histories and files, to get data from doctors and nurses and from the physical examination of patients, and to obtain samples (urine and stool) from patients.

At the theoretical level, they learned various techniques but did not apply them. Among those they did apply were urinalysis, stool analysis, and sugar analysis; techniques for diagnosing meningitis and tuberculosis; statistical techniques such as correlation and some experimental designs. In addition, they practiced some variants of techniques with which they were already familiar, such as electrophoresis, cell cultivation, ultracentrifuging, differential precipitation, and gas chromatography. They also learned to diagnose the presence of certain viruses in patients; they tried to produce antibodies for given viruses.

Each student, along with his or her teacher/adviser, had to plan the thesis research, design it, develop a protocol, and submit it to the

Internal Council for approval. Likewise, each student learned different techniques, depending on his or her research (e.g., enzyme purification, gene cloning, ways of determining the size of the trypanosome genome). Students also became familiar with some demographic parameters used to measure populations.

Third Cohort

The fourth year was devoted to studying and researching immunology (viruses, plasmids, antigens, and antibodies) and microbiology (bacteria genetics). The students also studied human pathology.

The early months were devoted to biomedicine. The professors requested that the students choose a research project with one of the researchers, for which purpose they had informal meetings in which the researchers explained their work to the students. Likewise, each professor taught a three-week theory course in her or his field of expertise. There were two or three weekly seminars in which the students discussed scientific articles that were to have been studied beforehand. They were given much reading material.

The pathology course lasted four months and was taught at the Neurology Hospital. The students learned clinical pathology at a theoretical level, however, since it was not possible to integrate them into the professor's experimental work, they took the course along with students from the School of Medicine. There was a total of fifty students in the course.

In the last semester, each student had to choose a thesis adviser and define a research problem. In this way, the students gradually joined different laboratories where they conducted their own research projects, which led to the writing of the undergraduate thesis.

Appendix 3

The Students' Psychological Traits

In order to know the students better and, possibly, compare them with other population groups, we studied some personality and intelligence traits of the first four cohorts, using psychological tests. Since the program is highly selective, the population group available from each cohort was very small. We worked with a total of seventeen students, six male and eleven female. The first cohort had four students, three male and one female. The second cohort was also made up of four students, again three male and one female. The third cohort was made up of six women, and the fourth cohort was formed by three women. Owing to the small size of our sample, the interindividual and intercohort variability was high, and the results cannot be conclusive. Our fundamental interest was to discover some cognitive and behavioral traits that could characterize science students. From our data we were able to hypothesize the existence of certain characterological and thought tendencies that may characterize science students and distinguish them from other professions.

Instruments

For this research, Raymond Cattell's sixteen-factor personality test (16PF), form A, and David Wechsler's adults' intelligence test (WAIS) were used (see Cattell and Eber 1962; Cattel et al. 1970; Wechsler 1955; Rapaport 1965; Karson and O'Dell 1987). Both tests were administered individually, by one of the authors at the beginning of the undergraduate program, in two sessions devoted to each test.

Statistical Treatment

Owing to the small number of students included in the sample, and the even smaller number of students per cohort, the statistical analysis had to be nonparametric.[1] We had to settle for a frequency analysis, ranking the students in relation to the different traits studied. We conducted Spearman correlation tests and discriminating analyses, which showed us the differences between the groups.

Student Traits

Each cohort presented a particular profile in the personality and intelligence traits studied, thereby differentiating it from the other cohorts, although there were also common tendencies and traits, as will be analyzed below. The differences are shown in Tables A.1, A.2, and A.3.

The students in the first cohort were characterized by their aggressiveness, competitiveness, and obstinacy (Factor E+, $\overline{X}$ = 9.5). They came across as rebellious, liked to experiment and to analyze stimuli from the medium, and preferred what was new over what was familiar to them; they were inquisitive and skeptical, trusting only what they themselves had verified. They proved reluctant to accept moralistic attitudes and were tolerant vis-à-vis ambiguity and change (Factor Q_1 +, $\overline{X}$ = 8.25). They were very imaginative, seeking to produce their own ideas—ideas that came from them. As they liked to fantasize, they would toy with strings of ideas in theoretical, aesthetic, or artistic matters (Factor M+, $\overline{X}$ = 8). They also proved to be individualistic and self-sufficient, tending to use many of their own resources and seeking stimuli for their own ideas, which they preferred over those of others (Factor Q_2 =, $\overline{X}$ = 7.25).

We observed a tendency in most (three) of the students of the first cohort to be reserved and to isolate or distance themselves from the social group. Generally they were skeptical, critical, and somewhat unemotional (A−). They showed good internal control, allowing them-

---

1. We thank Drs. Isabel Reyes Lagunes and Federico O'Reilly for their advice on statistics.

selves to handle the stimuli of the medium adequately (Factor C+), and they tended to be guided by their own needs more than by the demands of their social group (Factor Q3−). (See Tables A.1 and A.2.)

The second cohort's outstanding traits were a high degree of imaginativeness and fantasizing; they reflected a high degree of self-immersion for producing ideas, both in theoretical and in artistic matters (Factor M+, $\overline{X} = 8$). They showed laxness in their self-control, neglecting social rules in order to pursue their own needs instead. This trait was more pronounced in this cohort than in the first (Factor Q_3−, $\overline{X} = 3.5$). Factor C+ ($\overline{X} = 7$) also stood out, indicating a high degree of ego strength, which allowed them to control their emotions and focus on a project until they had finished it. Finally, they were characterized by having analytical, experimental minds, and they enjoyed toying with their own ideas (Factor $Q_1 =$, $\overline{X} = 7$). Factor B, which denotes intelligence, was not taken into account in this analysis; we considered it unnecessary because we were also using Wechsler's intelligence test.

The third cohort's outstanding traits were its experimenting, critical, analytical nature vis-à-vis the stimuli of the medium. They were tolerant regarding ambiguity and change (Factor $Q_1 + \overline{X} = 8.16$). It is interesting to note that the most outstanding students in the cohort scored higher for this trait and, by contrast, the least outstanding students had average scores. Moreover, the group in general showed predominantly obsessive traits. Another characteristic trait of this cohort was its high degree of imaginativeness and fantasizing ($M + \overline{X} = 7.66$). They showed very creative thinking and sought alternatives for a problem or idea. They enjoyed producing ideas, and through this they motivated themselves and attempted to be individualistic and original, preferring their own decisions. There was a tendency across the group, to an average or outstanding degree, to be aggressive and competitive (Factor E+).

The students of the fourth cohort were characterized by their suspiciousness, apprehensiveness, and distrust (Factor L+, $\overline{X} = 9.33$); they were aggressive and competitive, stubborn in asserting their own ideas: E+ ($\overline{X} = 7$). They showed an imaginative (Factor M+ $\overline{X} = 7$), analytical, and experimental (Factor $Q_1 + \overline{X} = 7.66$) mentality. Self-motivation predominated, as they were guided by their own ideas, even if this meant infringing upon social rules (Factor $Q_3 - \overline{X} = 3$). They even showed a certain social awkwardness, because of their simplicity and naïveté (Factor $N - \overline{X} = 3.66$). Factor $Q_1 +$ stood out to a large degree in two of the students, but another student scored low for the same factor. For this reason, even though the average is high, we cannot characterize the group as a whole; we can only point out that the majority tended to be rebellious, experimental, and analytical (Table A.2).

We obtained a profile of the students' personality traits by averaging the scores of the different subtests for the total population group, as can be seen in Table A.3. In the overall profile of the students in this undergraduate program, the outstanding factors were Q_1+ ($\bar{X} = 7.82$), $M+$ ($\bar{X} = 7.70$), and $E+$ ($\bar{X} = 7.41$), which indicate an experimental attitude, a high degree of imaginativeness and fantasizing, as well as rebelliousness, aggressiveness, and competitiveness.

As has been shown throughout this study, although distinctive traits do exist among the cohorts and a dispersion was observed in the scores, the scores tended to follow a similar pattern for certain traits. (See Fig. A.1.) The most characteristic aspect of these science students was their particular manner of perceiving the world and themselves, based on the internal stimuli they toyed with by fantasizing, thinking, and imagining. Overall, the sample did not present outstanding traits of dependence or submission to the group, nor did the students show tendencies of isolating themselves from each other (except in the case of the first cohort and some third-cohort students, although this did not become an outstanding trait). Nor did the traits of assuredness and self-confidence stand out.

On the Wechsler adults' intelligence scale, the students of all four cohorts reflected a high level of intelligence, although no student scored high enough to be classified as a genius. Five students had superior intelligence (average IQ was 121.4); eleven students had above-average intelligence (average IQ: 113); and only one student had average intelligence (IQ: 107). (See Table A.4.)

In analyzing the intelligence level by cohort, some cohorts were observed to have higher levels than others. This was particularly the case when comparing the individual results of each cohort, rather than the average intelligence quotients. Nevertheless, higher intelligence was not an outstanding characteristic of the cohorts. Indeed, among the cohorts that were perceived as "better" by the professors, one had a higher overall intelligence level, but for another cohort, considered the "best," most students had above-average intelligence, to our surprise.

As shown in Table A.4, the students of the first cohort had an average IQ of 118.5; three students showed superior intelligence (IQs of 120 and 122), and one showed above-average intelligence (IQ of 110). The second cohort had an average IQ of 117.7; most students had above-average intelligence (IQs of 116), and one student had superior intelligence (an IQ of 123). The third cohort had an average IQ of 113. Most students showed an above-average intelligence (IQs ranging from 110 to 115, with an average of 111.6). Only one student scored 120, which indicates superior intelligence. The fourth cohort had an average IQ of 111.33, with two students scoring above average (average IQ: 113.5) and one student scoring average (IQ: 107).

Thus, a high intelligence level, though desirable and highly sought by scientists who deal with students, does not in of itself characterize science students, at least in the sample studied. Rather, it was the constellation of personality traits, added to a good stock of intellectual knowledge, which stood out in these students. Hence, students with average or superior intelligence who used their aggressive impulses suitably, who were motivated by internal stimuli, who had a ludic attitude in their thought and imagination, and who defended their own ideas aggressively stood out.

Indeed, the scores of the different cohorts on the intelligence subscales are very similar to the profile we obtained, for which reason we will analyze them as a set. (For graphic information, see Figs. A.2 and A.3.) By analyzing the total intelligence profile of the four groups, we observe a higher level in the Verbal Quotient (IQ = 116) than in the Performance Quotient (IQ = 110.58), the Intelligence Quotient of the cohort profile being 115.11, which indicates an above-average intelligence (Tables A.5 and A.6). The difference between the two quotients is notable, as it points to a predominance of verbal intellectual functions over psychomotor functions among these students. This might explain their poor performance in laboratory work.

As for cognitive functions, students stood out in their ability to form concepts, developing conceptual generalizations (Similarities $\overline{X}$ = 14.05). A total of seven students reflected a level of abstract thought allowing them to make high-level generalizations. The remaining students engaged in functional thinking, allowing them to integrate relationships among disconnected elements, although they were not able to grasp the entire content. No case was observed of a student using concrete thought.

Although not all students showed a high level of abstraction, it is noteworthy that it is this subscale, related to abstraction ability and concept formation, which is the more indicative of the science students we analyzed. These data agree with the personality data given above, which stress the students' experimenting and inquisitive attitudes. We might be led to the conclusion that the students in this scientific undergraduate program were characterized by a ludic attitude in their thought and imagination, as well as by a high capacity for abstracting and forming concepts. Moreover, these concepts were greatly reinforced by the professors, since they are part of the scientific ideal.

Through their correct utilization of memory, all the students showed that they had an adequate stock and grasp of basic information; they reflected a high ability to integrate and apply information vis-à-vis the stimuli of the environment, and they could therefore solve problems adequately (Information $\bar{X} = 13.88$ on the profile, Vocabulary $\bar{X} = 12.64$, with minimal intercohort variations). Thus we concluded that these students handled themselves with greater ease—and correctly—by means of their conceptual skills, with the capacity for forming concepts predominating, for which they used the required internal information—or that of the medium—adequately.

Comparing individual output in the different subscales, we observed that six students presented scores on the Comprehensive subscale which were three or more units below those on the Vocabulary subscale; this may indicate a decrease in their judgment ability vis-à-vis a given situation. This, in turn, may point to a primacy of internal stimuli over external stimuli among the students, and to a low level of predominance of social strength over behavior. This result is interesting, for it corroborates the personality test data related to the predominance of internal stimuli in these individuals' behavior.

Something else that should be emphasized regarding the intelligence subscales has to do with the Block Design, which in the profile presents the fourth highest score ($\bar{X} = 12.11$) and makes the previous data more congruent. This test measures the formation of concepts through visual-motor organization. It comprises analytical functions (a model that must be "broken down" is presented for the subject to imitate) and synthetic functions (the subject reconstructs the model with his or her pieces). This subtest, then, is related to that of Similarities, and—at least among the science students included in this research project—the two subtests were in harmony.

Table A.4 shows the students' rankings in the WAIS subscales. All the students are clustered in the highest ranking for the subtests in Information, Similarities, and Vocabulary, the only exception being the last subtest, in which two students scored in a mean ranking. The ranking shows that, for the Block Design, students are clustered in the mean (eight students) and high (nine students) rankings, indicating a common direction vis-à-vis the analytical and synthetic concept-formation functions. These subtests indicate that the important characteristics of the science students we studied were their abilities in forming, abstracting, analyzing, synthesizing, and applying concepts.

Tables A.5 and A.6 show a comparison of intelligence profiles for each cohort of students as well as an overall profile based on the cohort profiles. The overall profile differences were small and statistically insignificant, indicating that the groups of students were relatively homogenous.

We conducted a comparison of the individual scores in both tests with the overall profiles (see Table A.7 and A.8). Once again, the observed differences were not statistically significant, for which reason the data should be taken as indicators of student "tendencies" that will have to be confirmed in subsequent studies.

We conducted correlation tests for the students' scores vis-à-vis the different personality factors, following the Spearman ranking test (see Table A.9). Factors E and Q2 showed a correlation of .6059, significant to .01. Thus, the higher the dominance, aggressiveness, and competitiveness (E+), the greater was the self-sufficiency, with the students preferring to make their own decisions. Factors A and Q3 had a negative correlation of −.6049, significant to .01, indicating that the most reserved, isolated, and critical (A−) students showed the greatest self-control. We observed a negative correlation of −.6048 between Factor I and H, significant to .01, implying that as the students showed more boldness they were more self-confident and pragmatic and accepted greater responsibility (I−). By contrast, the more emotionally sensitive they were (I+), the less bold and more shy, repressed, and inhibited (H−) they were. There also existed a negative correlation between Factors L and M of −.6383, significant to .01, indicating that the higher their self-confidence (L−), the higher their imagination and creativity (M+). We observed a positive correlation between Factors F and H of .5743, significant to .01. Peculiarly, the two traits are not distinctive to the group. Indeed, all the cohorts scored in a mean ranking. The relationship implies that the more impulsive they were (F+), the bolder they were (H+). Finally, there was a positive correlation between Factors

O and Q4 of .7739, significant to .001, reflecting that the more self-assured they were (O−), the less emotionally tense they were (Q4).

Table A.10 shows the correlation matrix vis-à-vis the WAIS test. There was a positive correlation of .7281, significant to .001, between the Information subtest and the Verbal Quotient, as well as between the Verbal Quotient and the Vocabulary subtest (positive correlation of .5806, significant to .01). Likewise, a positive correlation of .7534, significant to .001, was observed between the Picture Completion subtest and the Performance Quotient, as well as between the latter and the Block Design subtest (correlation of .5922, significant to .01). The Block Design subtest also correlated positively with the Intelligence Quotient (.7272, significant to .001). The Intelligence Quotient was also positively correlated with the Picture Arrangement subtest (.5741, significant to .01), with the Verbal Quotient (.7131, significant to .001), and with the Performance Quotient (.8429, significant to .001). We feel that these results point to the reliability and coherence of our data and the instruments we used.

Comparison Among Most- and Least-Socialized Students

In monitoring the cohorts over the years, through interviews with professors and students, we gradually determined which students were most easily socialized and those who had more difficulties. The criteria for distinguishing these individuals included the evaluation that professors made of the students as capable researchers, the degree of the students' integration into the research group, the students' involvement in research tasks, the students' evaluations of themselves as researchers—that is, whether they did or did not see themselves as part of the group—and, finally, the extent to which students joined the scientific group, as a part of the work team, pursuing graduate studies in their field.

Our intention was not to score the degree of socialization or integration into the group, but to attempt to determine whether there were personality or intelligence differences between the less-socialized students and the rest of the group. Our purpose was to analyze the possible differences, to find out if certain traits were more characteristic of the scientific group and, therefore, made them stand out among the candidates.

Thus we selected, from each cohort, the least socialized student; that is, the student who had the greatest difficulty integrating her- or himself into the group of scientists. We conducted this selection at the end of the research project, when the fourth-cohort students were in the fourth year of the program and the previous cohorts had begun their graduate studies—or had changed their field of specialization. Nevertheless, we acknowledge that this grouping is arbitrary and does not follow criteria of academic quality among the students, since not even the least-socialized students were poor students.

We compared the personality and intelligence test scores of the group of least-socialized students with those of the rest of the group, distinguishing the intensity of traits by rank. Because of the small number of subjects (four students were compared with thirteen), it was difficult to conduct meaningful statistical tests (see Tables A.11 and A.12). These tables present the frequencies for the different rankings in the traits studied. As shown, the groups are similar in regard to the different intelligence subscales; that is, the least-socialized students did not have a significantly lower or different intelligence level. It was not intelligence that set them apart, but their personality traits, which determined a different manner of interacting with others. A comparison of the score frequencies for both the instruments used shows more clearly the similarity in intelligence subscales between the two groups (Table A.13).

On the personality test, though, certain differences can be seen in Factors A, C, G, H, I, Q2, and Q3. A tendency toward A+ (a higher degree of openly displaying affection) can be observed in the least-socialized group, whereas the most-socialized group tended toward A− (more reserved, aloof, critical, isolated). The least-socialized group reflected a greater tendency toward C− (emotional instability, weak ego), whereas the most-socialized students tended toward C+ (greater ego strength). The least-socialized students reflected a tendency toward G+, showing a tendency to conform to the group, to be moralistic and formal. The most-socialized students represented Factor H+ (bold, enterprising), whereas the least-socialized students showed intermediate-level or low scores in this trait and came across as more timid. An opposite tendency was also observed in Factor I, showing that the least-socialized students had higher sensitivity (I+), whereas the more-socialized students tended to come across as harsher and unsentimental; the latter did not govern themselves according to illusions and were more self-confident (I−).

Another clear tendency was for the students who were the least socialized vis-à-vis Q2+ to reflect a high degree of self-sufficiency, whereas the other group reflected a different orientation. Likewise, all the least-socialized students reflected a tendency toward Q3−, reflecting little self-control, whereas the other group presented the same tendency to a lesser degree (i.e., greater self-control, though within a low ranking).

We conducted a statistical comparative analysis in which we considered only some personality factors (Factors A, C, E, H, I M, O, Q1, Q2, and Q3) and the students' IQs. For each individual, the values of these variables were standardized vis-à-vis the student's cohort. The purpose of this standardization was to eliminate the possible influence of cohort variation.

We proceeded to see whether the a priori classification, after we had removed the least-socialized student from each cohort, could be verified through the data. Based on a variance analysis, a discriminating variance was constructed, using Fisher's linear discriminating functions. Thus, we combined the values of the variables studied with statistical weights, and we compared the two groups. This analysis separated the sample subjects perfectly, as shown in Figure A.4. The data confirm that our classification was not arbitrary and that the two groups were indeed distinguished from one another.

We correlated the variables and scores of the different groups, and we found that the variable which most distinguished the relatively socialized students was Factor I, which correlated negatively (−.14422); hence, whereas the stronger and more stringently self-controlling students were the most socialized, the least-socialized students were the most emotionally sensitive. Factor A also correlated negatively for the two groups (−.09229), indicating that the least-socialized group had more affective expression and that the most-socialized group was more reserved.

In this statistical analysis, the IQ appears once again as a nondiscriminating factor between the two groups. In any case, it can be seen that the size of the correlations is too small to do reliability tests. Table A.14 shows the ordering of the variables correlated between the two groups.

The personality traits and modes of thinking that we have analyzed provide us with an idea of these students' distinctive traits; nevertheless, this study cannot be considered conclusive. A broader clinical study must be done, through projective and objective tests applied to larger populations, preferably at the beginning and end of training, in order to have a more accurate profile of science students. We hope that these data may serve as a basis for such a study. Our presentation of these data, notwithstanding their lack of statistical significance, is intended to allow such a study to be conducted.

The Students' Psychological Traits   193

Table A.1. Personality traits (16PF)

Cohorts	Subjects	Factors															
		A	B	C	E	F	G	H	I	L	M	N	O	Q1	Q2	Q3	Q4
I	1	4	4	7	10	3	6	6	2	4	9	6	4	6	8	5	5
	2	7	5	6	8	5	5	6	9	4	10	4	3	10	7	4	3
	3	3	6	7	10	5	6	5	7	8	7	6	5	10	8	3	5
	4	3	3	5	10	7	4	8	1	8	6	9	6	7	6	4	5
II	5	3	7	5	10	6	5	6	5	6	8	3	5	8	9	6	6
	6	7	7	7	6	8	5	8	7	6	6	9	6	6	6	1	5
	7	5	5	6	5	4	5	4	6	6	8	4	2	6	7	3	4
	8	6	9	10	6	4	9	8	4	4	10	4	3	8	2	4	1
III	9	4	5	7	8	7	3	7	3	6	6	3	4	10	8	6	2
	10	2	6	8	7	3	5	5	6	5	6	5	5	9	8	5	5
	11	6	6	7	6	6	4	5	6	8	9	5	7	10	4	5	5
	12	4	8	5	8	5	6	7	7	6	9	5	10	8	6	6	7
	13	7	7	4	6	6	7	6	8	8	6	6	7	6	7	1	7
	14	3	6	8	5	5	5	6	5	4	10	4	7	6	6	9	6
IV	15	5	6	4	9	2	5	5	8	10	6	5	7	4	9	2	9
	16	6	8	5	5	6	5	8	3	8	9	4	6	10	5	2	4
	17	4	4	6	7	7	4	7	4	10	6	2	6	9	6	5	5

NOTE: Standardized scores.

Table A.2. Ranking of personality traits (16PF)

Cohorts	Subjects	A l m h	B l m h	C l m h	E l m h	F l m h	G l m h	H l m h	I l m h	L l m h	M l m h	N l m h	O l m h	Q1 l m h	Q2 l m h	Q3 l m h	Q4 l m h
I	1	4	4	7	10 3	6	6	6	2	4	9	6	4	6	8	5	5
	2	7	5	6	8	5	5	6	9	4	10 4	3	3	10	7	4	3
	3	3	6		10	5	6	5	7		7	6	5	10	8	3	5
	4	3	3	5	10	7 4			8 1	8	6	9	6	7	6	4	4.5
	X̄	4.25	4.50	6.25	9.5	5	5.25	6.25	4.75	6	8	6.25	4.5	8.25	7.25	4	4.5
II	5	3	7	5	10	6	5	6	5	6	8	3	5	6	9	6	6
	6	7	7	7	6	8	5	8 2	8 2	8	6	9	6	6	6	1	5
	7	5	5	6	5 4	4	5	4	6	8	8 4	2		6	7 3	3	4
	8	6	9	10	6 4		9	8 4	8 4	4	10 4	3		8 2	4		1
	X̄	5.25	7	7	6.75	5.5	6	6.5	4.25	6.5	8	5	4	7	5	3.5	4
III	9	4	5	7	8	7 3	5	7 3	6	6	6	5 3	4	10	8	6	2
	10	2	6	8	7 3	6	5	5	6	5	6	5	5	9	8	5	5
	11	6		7	6	6 4	6	5	6	8	9	5		10 4	5	5	5
	12	4	8 5		8	5	6	7	7	8	9	5	7	8	6	6	7
	13		7 4		6	6		6 7	8	8	6	6	10	6	7 1		7
	14	3	6	8	5	5	5	6	5	4	10 4		7	6	6	9	6
	X̄	4.33	6.33	6.5	6.66	5.33	5	6	5.83	6.16	7.66	4.66	6.66	8.16	6.5	5.33	5.33
IV	15	5	6 4		9 2	5	5	5		8	6	5		7 4	9 2	9 2	9
	16	6	8	5	5	6	5	8 3	8 3	8	9 4	9 4	6	10	5	5 2	
	17	4 4		6	7	7 4		7 4	7 4	10	6 2	2	6	9	6		4 5
	X̄	5.0	6	5	7	5	4.66	6.29	5	9.35	3.66	6.33	7.66	6.66	3	6	5

NOTE: l = low, score 1–4; m = medium, score 5–6; h = high, score 7–10.

Table A.3. Comparison of personality profiles (16PF), by cohort

	A	B	C	E	F	G	H	I	L	M	N	O	Q1	Q2	Q3	Q4
Total profile	x = 4.64	6	6.29	7.41	5.23	5.23	6.29	5.05	6.76	7.70	4.94	5.47	7.82	6.58	4.17	4.94
I	x = 4.25	4.50	6.25	9.5	5	5.25	6.25	4.75	6	8	6.25	4.5	8.25	7.25	4	4.5
II	x = 5.25	7	7	6.75	5.5	6	6.5	4.25	6.5	8	5	4	7	6	3.5	4
III	x = 4.33	6.33	6.5	6.66	5.33	5	6	5.83	6.16	7.66	4.66	6.66	8.16	6.5	5.33	5.33
IV	x = 5.00	6	5	7	5	4.66	6.66	5	9.33	7	3.66	6.33	7.66	6.66	3	6
								Differences from Total Profile								
I	−.39	−1.5	−.04	2.09	−.23	.02	−.04	−.30	−.76	.30	1.31	−.97	.43	.67	−.17	−.44
II	.61	1	.71	−.66	.77	.77	.21	−.80	−.26	.30	.06	−1.97	−.82	−.58	−.67	−.94
III	−.31	.33	.21	−.75	.10	−.23	−.29	.78	−.6	−.04	−.28	1.19	.34	−.08	1.16	.39
IV	.36	0	−1.29	−.41	−.23	−.57	.37	−.05	2.57	−.70	−1.28	.86	−.16	.08	−1.17	1.06

Table A.4. Intelligence traits (WAIS)

Cohorts	Subjects	Verbal						Performance					Verbal IQ	Performance IQ	IQ
		Infor-mation	Compre-hension	Arith-metic	Simi-larities	Digit Span	Vocab-ulary	Digit Symbol	Picture Completion	Block Design	Picture Arrangement	Object Assembly			
I	1	13	9	12	12	15	12	15	16	16	13	10	115	127	122
	2	14	16	10	15	12	14	11	12	15	14	10	121	115	120
	3	12	9	14	13	9	12	15	9	11	12	11	109	99	110
	4	15	11	11	12	15	14	10	12	15	14	13	121	119	122
II	5	14	15	13	12	10	13	10	11	13	10	12	119	109	116
	6	14	12	9	17	15	11	13	11	10	10	12	120	109	116
	7	17	12	12	18	11	13	10	13	16	11	12	125	116	123
	8	14	12	10	13	9	12	13	14	10	13	17	110	121	116
III	9	13	14	12	12	10	13	10	10	12	9	12	114	103	110
	10	14	13	12	15	6	14	12	13	14	10	9	116	111	115
	11	15	9	14	16	15	13	16	11	10	11	11	124	113	120
	12	13	13	13	15	7	12	10	12	12	9	14	113	108	112
	13	13	13	9	16	11	13	11	9	10	10	10	117	101	111
	14	13	9	12	13	12	12	12	11	9	9	11	113	103	110
IV	15	17	10	14	13	7	13	10	11	10	11	11	114	103	110
	16	12	7	11	13	14	11	11	14	9	9	11	108	104	107
	17	13	9	11	14	11	11	13	12	14	9	16	113	119	117

NOTE: IQ of 90–109 = average intelligence; 110–119 = above-average; 120–129 = superior; 130 or higher = very superior. Scores of 1–6 = 5–35%; 7–13 = 40–70%; 14–19 = 75–100%.

Table A.5. Ranking of intelligence traits (WAIS)

Cohorts	Subjects	Information			Comprehension			Arithmetic			Similarities			Digit Span			Vocabulary			Digit Symbol			Picture Completion			Block Design			Picture Arrangement			Object assembly			Verbal IQ			Performance IQ			IQ		
		l	m	h	l	m	h	l	m	h	l	m	h	l	m	h	l	m	h	l	m	h	l	m	h	l	m	h	l	m	h	l	m	h	l	m	h	l	m	h	l	m	h
I	1			13		9				12			12			15			12			15			16			16			13		10			115				127			122
I	2			14			16		10				15			12			14		11				12			15			14		10				121		115				120
I	3			12		9				14			13		9				12			15		9			11				12		11		109			99				110	
I	4			15		11			11				12			15			14		10				12			15			14			13			121			119			122
II	5			14			15			13			12		10				13		10			11				13		10				12		119			109			116	
II	6			14			12		9				17			15	11				13			11			10			10				12			120		109			116	
II	7			17			12			12			18		11				13		10				13			16		11				12			125			116			123
II	8			14			12		10				13		9				12			13			14		10				13			17		110				121		116	
III	9			13			14		12				12		10				13		10			10				12		9				12		114			103			110	
III	10			14			13			12			15	6					14		12				13			14			10		9			116			111			115	
III	11			15		9				14			16			15			13			16		11			10			11		11				124			113			120
III	12			13			13			13		7	15						12		10				12			12		9				14		113		108				112	
III	13			13			13		9				16			11			13		11			9			10			10		10				117		101				111	
III	14			13		9				12			13			12			12			12		11			9			9		11			113			103			110	
IV	15			17		10				14			13	7					13		10			11			10			11		11			114			103			110	
IV	16			12	7				9			11				13			14		11			11				14		9			11		108			104			107		
IV	17			13		9			11				14		11				13			13			12			14		9			9			113			119			117

NOTE: l = low (score 0–8.5, IQ 90–109); m = medium (score 8.6–11.5, IQ 110–119); h = high (score 11.6 or higher, IQ 120–129).

Table A.6. Comparison of intelligence profiles (WAIS), by cohort

		Infor-mation	Compre-hension	Arith-metic	Simi-larities	Digit Span	Vocab-ulary	Digit Symbol	Picture Completion	Block Design	Picture Arrangement	Object Assembly	Verbal IQ	Performance IQ	IQ
Total profile	x =	13.88	11.35	11.70	14.05	11.11	12.64	11.88	11.82	12.11	10.82	11.88	116	110.58	115.11
I	x =	13.50	11.25	12	13	12.75	13	12.75	12.25	14.25	13.25	11	115.50	115.	118.50
II	x =	14.75	12.75	11	15	11.25	12.25	11.50	12.25	12.25	11	13.25	118.7	113.7	117.7
III	x =	13.50	11.83	12	14.5	10.16	12.83	11.83	11	11.16	9.66	11.16	116.16	106.5	113
IV	x =	14	8.66	12	13.33	10.66	12.33	11.33	12.53	11	9.66	12.66	111.66	108.66	111.33
								Differences from Total Profile							
I		−.38	−.10	.30	−1.05	1.64	.36	.87	.43	2.14	2.43	−.88	.50	4.42	3.39
II		.87	1.40	−.70	.95	.14	−.39	−.38	.43	.14	.18	1.37	2.70	3.12	2.59
III		−.38	0.48	.30	.45	−.95	.19	−.05	−.82	−.95	−1.16	−.72	.16	−4.08	2.11
IV		.12	−2.69	.30	−.72	−.45	−.31	−.55	.71	−1.11	−1.16	.78	−4.34	−1.92	−3.78

Table A.7. Comparison of the total personality profile (16PF): Differences from the cohort means

Cohorts	Subjects	A x = 4.64	B 6	C 6.29	E 7.41	F 5.23	G 5.23	H 6.29	I 5.05	L 6.76	M 7.70	N 4.94	O 5.47	Q1 7.82	Q2 6.58	Q3 4.17	Q4 4.94
I	1	.64	-2	.71	2.59	-2.23	.77	-.29	-3.05	-2.76	1.30	1.06	-1.47	-1.82	1.42	.83	.06
	2	2.36	-1	-.29	.59	-.23	-.23	-.29	3.95	-2.76	2.30	-.94	-2.47	2.18	.42	-.17	-1.94
	3	-1.64	0	.71	2.59	-.23	.77	-1.29	1.95	1.24	.70	1.06	-.47	2.18	1.42	-.17	.06
	4	-1.64	-3	-1.29	2.59	1.77	-1.23	1.71	-4.05	1.24	-1.70	4.06	.53	-.82	-.58	-.17	.06
II	5	-1.64	1	-1.29	2.59	.77	-.23	-.29	-.05	-.76	.30	-1.94	-.47	.18	2.42	1.83	1.06
	6	2.36	1	.71	-1.41	2.77	-.23	1.71	-3.05	1.24	-1.70	4.06	.53	-1.82	-.58	-3.17	.06
	7	1.64	-1	-.29	-2.41	-1.23	-.23	-2.29	.95	1.24	.30	-.94	-3.47	-1.82	.42	-1.17	-.94
	8	1.36	3	3.71	-1.41	-1.23	3.77	1.71	-1.05	-2.76	2.30	-.94	-2.47	.18	-4.58	-1.17	-3.94
III	9	-.64	1	.71	.59	1.77	-2.23	.71	-2.05	-.76	-1.70	-1.94	-1.47	2.18	1.42	1.83	-2.94
	10	-2.64	0	1.71	-.41	-2.23	-.23	-1.29	.95	-1.76	-1.70	.06	-.47	1.18	1.42	.83	.06
	11	1.36	0	.71	1.41	.77	-1.23	-1.29	.95	1.24	1.30	.06	1.53	2.18	-2.58	.83	.06
	12	-.64	2	-1.29	.59	-.23	.77	.71	1.95	-.76	1.30	.06	4.53	.18	-.58	1.83	2.06
	13	2.36	1	-2.29	-1.41	.77	1.77	-.29	2.95	1.24	-1.70	1.06	1.53	-1.82	.42	-3.17	2.06
	14	-1.36	0	1.71	2.41	.23	-.23	-.29	-.05	-2.76	2.30	-.94	1.53	-1.82	-.58	4.83	1.06
IV	15	.36	0	-2.29	1.59	3.23	-.23	-1.29	2.95	3.24	-1.70	.06	1.53	-3.82	2.42	-2.17	4.06
	16	1.36	2	-1.29	-2.41	.77	-.23	1.71	-2.05	1.24	1.30	-.94	.53	2.18	-1.58	-2.17	-.94
	17	-.64	-2	-.29	-.41	1.77	-1.23	.71	-1.05	3.24	-1.70	-2.94	.53	1.18	-.58	.83	.06

Table A.8. Comparison of the total intelligence profile (WAIS): Differences from the cohort means

Cohorts	Subjects	Information x̄ = 13.88	Comprehension 11.35	Arithmetic 11.70	Similarities 14.05	Digit Span 11.11	Vocabulary 12.64	Digit Symbol 11.88	Picture Completion 11.82	Block Design 12.11	Picture Arrangement 10.82	Object Assembly 11.83	Verbal IQ 116	Performance IQ 110.58	IQ 115.11
I	1	-.88	-2.35	.30	-2.05	3.89	-.64	3.12	4.18	3.89	2.18	-1.88	-1	16.42	6.89
	2	.12	4.65	-1.70	.95	.89	1.36	-.88	.18	2.89	3.18	-1.88	5	4.42	4.89
	3	-1.88	-2.35	2.30	1.05	-2.11	-.64	3.12	-.82	-1.11	1.18	-.88	-7	11.58	-5.11
	4	1.12	-.35	-.70	-2.05	3.89	1.36	-1.88	.18	2.89	3.18	1.12	5	8.42	6.89
II	5	.12	3.65	1.30	-2.05	-1.11	.36	-1.88	-.82	.89	-.82	.12	3	-1.58	.89
	6	.12	.65	-2.70	2.95	3.89	-1.64	1.12	-.82	-2.11	-.82	.12	4	-1.58	.89
	7	3.12	.65	.30	3.95	-.11	.36	-1.88	1.18	3.89	.18	.12	9	5.42	7.89
III	8	.12	.65	-1.70	-1.05	-2.11	-.64	1.12	2.18	-2.11	2.18	5.12	-6	10.42	.89
	9	-.88	2.65	.30	-2.05	-1.11	.36	-1.88	-1.82	.11	-1.82	.12	-2	-7.58	-5.11
	10	.12	1.65	.30	.95	-5.11	1.36	.12	1.18	1.89	-.82	-2.88	0	.42	-.11
	11	1.12	-2.35	2.30	1.95	3.89	.36	4.12	-.82	-2.11	.18	-.88	8	2.42	4.89
	12	-.88	1.65	1.30	.95	-4.11	-.64	-1.88	.18	.11	-1.82	2.12	-3	-2.58	-3.11
	13	-.88	1.65	-2.70	1.95	-.11	.36	-.88	-2.82	-2.11	-.82	-1.88	1	9.58	-4.11
	14	-.88	-2.35	.30	-1.05	.89	-.64	.12	-.82	-3.11	-1.82	-.88	-3	-7.58	-5.11
IV	15	3.12	-1.35	2.30	-1.05	-4.11	.36	-1.88	-.82	-2.11	.18	-.88	-2	-7.58	-5.11
	16	-1.88	-4.35	-.70	-1.05	2.89	-1.64	-.88	2.18	-3.11	-1.82	-.88	-8	-6.58	-8.11
	17	-.88	2.35	-.70	-.05	-.11	.36	1.12	.18	1.89	-1.82	4.12	-3	8.42	1.89

Table A.9. Correlations matrix for 16PF

	A	B	C	E	F	G	H	I	L	M	N	O	Q1	Q2	Q3	Q4
A	1.0000	0.2994	-0.2061	-0.4454	0.1573	0.1788	0.2021	0.1783	0.1580	0.1685	0.0759	0.0013	-0.0494	-0.3871	-0.6049*	-0.2332
B	0.2994	1.0000	-0.0294	-0.3598	-0.0565	0.5508	0.2330	0.1973	-0.0993	0.2699	-0.0006	0.2874	0.0083	-0.2865	-0.1803	0.1633
C	-0.2061	-0.0294	1.0000	-0.2221	-0.1631	0.0299	-0.0271	-0.2775	-0.5523	0.3102	-0.0737	-0.3854	0.2160	-0.2210	0.3429	-0.4770
E	-0.4454	-0.3598	-0.2221	1.0000	-0.0704	-0.0060	-0.0577	-0.0082	-0.0372	-0.2501	0.2278	-0.1097	0.0642	0.6059*	0.1966	0.2202
F	0.1573	-0.0565	-0.1631	-0.0704	1.0000	-0.5196	-0.5743*	-0.4474	0.3506	-0.3454	-0.0446	0.2158	0.2787	-0.3323	-0.0221	-0.1060
G	0.1788	0.5508	0.0299	-0.0060	-0.5196	1.0000	-0.0428	0.3188	-0.3329	0.3493	0.3154	-0.0496	-0.3018	0.0273	-0.2748	0.1864
H	0.2021	0.2330	-0.0271	-0.0577	-0.5743*	-0.0428	1.0000	-0.6408*	-0.0818	0.0211	-0.0161	0.0755	0.0836	0.5300	-0.0306	-0.2623
I	0.1783	0.1973	-0.2775	-0.0082	-0.4474	0.3188	-0.6408*	1.0000	0.0536	0.1239	-0.0473	0.1736	0.0526	0.2855	-0.1268	0.3425
L	0.1580	-0.0993	-0.5523	-0.0372	0.3506	-0.3329	-0.0818	0.0536	1.0000	-0.6383*	0.1463	0.3963	-0.0904	-0.0105	-0.5243	0.2830
M	0.1685	0.2699	0.3102	-0.2501	-0.3454	0.3493	0.0211	0.1239	-0.6383*	1.0000	-0.2565	-0.1662	0.1794	-0.4012	0.3080	-0.2583
N	0.0759	-0.0006	-0.0737	0.2278	-0.0446	0.3154	-0.0161	-0.0473	0.1463	-0.2565	1.0000	0.2936	-0.3527	-0.0026	-0.4749	0.3279
O	0.0013	0.2874	-0.3854	-0.1097	0.2158	-0.0496	0.0755	0.1736	0.3963	-0.1662	0.2936	1.0000	-0.2159	-0.2311	0.0246	0.7739†
Q1	-0.0494	0.0083	0.2160	0.0642	0.2787	-0.3018	0.0836	0.0526	-0.0904	0.1794	-0.3527	-0.2159	1.0000	-0.1823	0.2286	-0.5275
Q2	-0.3871	-0.2865	-0.2210	0.6059*	-0.3323	0.0273	0.5300	0.2855	-0.0105	-0.4012	-0.0026	-0.2311	-0.1823	1.0000	0.0635	0.3024
Q3	-0.6049*	-0.1803	0.3429	0.1966	-0.0221	-0.2748	-0.0306	-0.1268	-0.5243	0.3080	-0.4749	0.0246	0.2286	0.0635	1.0000	0.0561
Q4	-0.2332	0.1633	-0.4770	0.2202	-0.1060	0.1864	-0.2623	0.3425	0.2830	-0.2583	0.3279	0.7739†	-0.5275	0.3024	0.0561	1.0000

NOTE: N = 17.
*P = .01.
†P = .001.

Table A.10. Correlations matrix for WAIS

	Infor-mation	Compre-hension	Arith-metic	Simi-larities	Digit Span	Vocab-ulary	Digit Symbol	Picture Completion	Block Design	Picture Arrangement	Object Assembly	Verbal IQ	Performance IQ	IQ
	1.0000	0.2653	0.0946	0.2541	0.0373	0.5245	−0.2721	0.1110	0.2647	0.4792	0.1117	0.7281†	0.3969	0.5554
	0.2653	1.0000	−0.1996	0.1147	−0.3758	0.4770	−0.5260	−0.1795	0.3306	0.0574	0.0140	0.3993	−0.0376	0.1247
	0.0946	−0.1996	1.0000	−0.2197	−0.3607	0.0375	−0.0501	−0.2219	0.0178	−0.0492	−0.0753	0.0859	−0.2709	−0.1697
	0.2541	0.1147	−0.2197	1.0000	0.0285	0.0185	0.1688	−0.0537	−0.0833	−0.0833	−0.1267	0.3835	−0.0289	0.1986
	0.0373	−0.3758	−0.3607	0.0285	1.0000	−0.1456	0.3098	0.1158	0.0214	0.2066	−0.0829	0.4018	0.3389	0.4355
	0.5245	0.4770	0.0375	0.0185	−0.1456	1.0000	−0.3494	−0.0758	0.5476	0.3190	−0.1946	0.5806*	0.2093	0.3899
	−0.2721	−0.5260	−0.0501	0.1688	0.3098	−0.3494	1.0000	0.0663	−0.1889	0.1792	−0.2312	−0.1765	0.2063	0.1159
	0.1110	−0.1795	−0.2219	−0.0537	0.1158	−0.0758	0.0663	1.0000	0.3666	0.2041	0.0824	−0.0628	0.7534†	0.4201
	0.2647	0.3306	0.0178	−0.0833	0.0214	0.5476	−0.1889	0.3666	1.0000	0.4163	0.0102	0.4704	0.5922*	0.7272†
	0.4792	0.0574	−0.0492	−0.0833	0.2066	0.3190	0.1792	0.2041	0.4163	1.0000	−0.1613	0.4089	0.4623	0.5741*
	0.1117	0.0140	−0.0753	−0.1267	−0.0829	−0.1946	−0.2312	0.0824	0.0102	0.1613	1.0000	−0.1627	0.2789	0.1536
	0.7281†	0.3993	−0.0859	0.3835	0.4018	0.5806*	−0.1765	−0.0628	0.4704	0.4089	−0.1627	1.0000	0.3340	0.7131†
	0.3969	−0.0376	−0.2709	−0.0289	0.3389	0.2093	0.2063	0.7534†	0.5922*	0.4623	0.2789	0.3340	1.0000	0.8429†
	0.5554	0.1247	−0.1697	0.1986	0.4355	0.3899	0.1159	0.4201	0.7272†	0.5741*	0.1536	0.7131†	0.8429†	1.0000

NOTE: N = 17.
*P = .01.
†P = .001.

Table A.11. Comparison of most- and least-socialized students: Personality (16PF)

Subjects	A l	A m	A h	B l	B m	B h	C l	C m	C h	E l	E m	E h	F l	F m	F h	G l	G m	G h	H l	H m	H h	I l	I m	I h	L l	L m	L h	M l	M m	M h	N l	N m	N h	O l	O m	O h	Q1 l	Q1 m	Q1 h	Q2 l	Q2 m	Q2 h	Q3 l	Q3 m	Q3 h	Q4 l	Q4 m	Q4 h
2		7			5			6				8		5			5			6				9					10	4				3					10		7	4		7	4			3	
7	5				5			6			5		4				5		4				6				8		8	4					2			6			7	3		7	3			4	
13		7				7 4					6			6				7		6				8					6			6								7			7							7
15	5				6 4						6			6			5			5				8		10			6				5						7 4		9	2							9
(Frequencies)	(2)(2)			(3) (1)(2) (2)						(2)(2) (2) (2)			9 (2)			(3)(1)			(1)(3)			(1)(3) (1)					(3)	(2) (2)(2)(2)					(2)			(2) (1) (2)(11)			(4) (4)						(2)		
1	4				4				7					3			6			6			2				4			9 6			6			4		6			8	5		8	5				
3	3				6				7		5		10				6			5				7						7		6							10		8	3		8	3				
4	3				3				5		10			5			6				8 1						8		6				9						7			6	4		6	4				
5	3				7	5					10		6				5			6					6		8			8 3						5			8			9			9				6	
6		7			7				7		6			8			5				8 2			5					6				9			6			6		6		1		6	1				5
8		6				9			10		6			4				9			8 4				4				10 4						3			6		8	2		8	2			4		1
9	4				5			7				8		7	3					7 3							6		6	3		5				4		10			8	6		8	6				2
10	2				6			8			7			5			5		5				6				8		6				5		5			9			8	5		8	5			5	
11		6			6			7		6			7 3				5		5				6			6			9			5						10			4		5		5			5	
12	4				8			5			8			5			6			7			7				8		9			5			7			8			6			6					7
14	3				6			8			5			5			5			6			5			4			6			5			10			7			6			6				6	
16		6			8			5			5			6			5			8 3							8		10	4					6			10			6			6			4		
17	4				4			6		7			7 4				5			7 4							10		6	2					6			9			6			6				5	
(Frequencies)	(9)(3)(1)(3)(5)			(5)(1) (4) (8) (0) (5)(8)						(3) (6) (4)(4)(8)(1)						(0)(6)(7) (7) (4)(2)			(3) (4)(6)(0)(5)			(8)(6)(5) (2)			(3) (7)(3)			(0)(3)(10)						(2)(6)(5)(5)(2)(1)						(3)(9)(1)		

NOTE: l = low, score 1–4; m = medium, score 5–6; h = high, score 7–10.

Table A.12. Comparison of most- and least-socialized students: Intelligence (WAIS)

Subjects	Information			Comprehension			Arithmetic			Similarities			Digit Span			Vocabulary			Digit Symbol			Picture Completion			Block Design			Picture Arrangement			Object assembly			Verbal IQ			Performance IQ			IQ		
	l	m	h	l	m	h	l	m	h	l	m	h	l	m	h	l	m	h	l	m	h	l	m	h	l	m	h	l	m	h	l	m	h	l	m	h	l	m	h	l	m	h
2	14			16			10			15			12			14			11			12			15			14			10			121			115			120
7	17			12			12			18			11			13			10			13			16			11			12			125			116			123
13	13			13			9			16			11			13			11			9			10			10			10			117			101			111
15	17			10			14			13	7		13			13			10			11			10			11			11			114			103			110
(Frequencies)	(4)			(1) (3)			(2)(2)			(4)(1) (2) (1)			(4)			(4)			(2)(2)			(2)(2)			(3) (1)			(3) (1)			(2) (2)			(2) (2)			(2)
1	13			9			12			12			15			12			15			16			16			13			10			115			127			122
3	12			9			14			13			9			12			15			9			11			12			11 109					99			110
4	15			11			11			12			15			14			10			12			15			14			13			121			119			122
5	14			15			13			12			10			13			10			11			13			10			12			119			109			116
6	14			12			9			17			15			11			13			11			10			10			12			120			109			116
8	14			12			10			13			9			12			13			14			10			13			17 110					121			116
9	13			14			12			12			10			13			10			10			12			9			12			114			103			110
10	14			13			12			15	6		14			12			13			14			10			9			116			111			115
11	15			9			14			16			15			13			16			11			10			11			11			124			113			120
12	13			13			13			15	7		12			10			12			12			9			14			113			108			112
14	13			9			12			13			12			12			12			11			9			11			113			103			110
16	12	7								13			14			11			11			14			9			9			11			108			104			107
17	13			9			11			14			11			13			13			12			14			9			16			113			119			117
(Frequencies)	(13)(1)(6)			(6)			(5)(8)			(13)(2)(4)			(7)			(2)(11)			(5) (8)			(6) (7)			(6) (7)			(9) (4)			(5) (8)			(3) (7)			(3) (7) (4)			(2)			(1) (9) (3)

NOTE: l = low (score 0–8.5, IQ 90–109); m = medium (score 8.6–11.5, IQ 110–119); h = high (score 11.6 or higher, IQ 120–129).

Table A.13. Frequency comparison of least-socialized students against rest of group

	A			B			C			E			F			G			H			I			L			M			N			O			Q1			Q2			Q3			Q4		
	l	m	h	l	m	h	l	m	h	l	m	h	l	m	h	l	m	h	l	m	h	l	m	h	l	m	h	l	m	h	l	m	h	l	m	h	l	m	h	l	m	h	l	m	h	l	m	h
Least socialized		2	2	3	1			2	2	2	2	2	2	2	2	3	1		1	1	3	1	3	1		3		2	2	2	2	2	2	2	2	3	1	2	1		4	4		5	5	7	1		3	9	1

(16PF — values for Socialized row: 9 3 1 3 5 5 1 4 8 5 8 3 6 4 4 8 1 0 6 7 7 4 2 3 4 6 5 8 6 5 2 3 7 3 3 10 2 6 5 5 7 1 3 9 1)

	Information			Comprehension			Arithmetic			Similarities			Digit Span			Vocabulary			Digit Symbol			Picture Completion			Block Design			Picture Arrangement			Object assembly			Verbal IQ			Performance IQ			IQ		
	l	m	h	l	m	h	l	m	h	l	m	h	l	m	h	l	m	h	l	m	h	l	m	h	l	m	h	l	m	h	l	m	h	l	m	h	l	m	h	l	m	h
Least socialized		4		1	3			2	2	4			4	1	2			4		4	4		2	2		2	2	3	1		3	1			2	2		2	2		2	2
Socialized	13			1	6	6	5	8			13		2	4	7	2	11		5	8		6	7		6	7		9	4		5	8		3	7	3	7	4	2	1	9	3

NOTE: The least-socialized students are Subjects 2, 7, 13, and 15 (see Tables A.11 and A.12). l = low; m = medium; h = high.

Table A.14. Correlations between the most- and least-socialized groups for some personality and intelligence variables

Variables	Correlation
I	−.14422
A	−.09229
H	.08882
C	.08497
Q3	.07919
Q1	.06702
Q2	−.05298
O	.03775
E	.03033
IQ	−.01890
M	.00931

NOTE: Variables are ordered according to the size of the correlation.

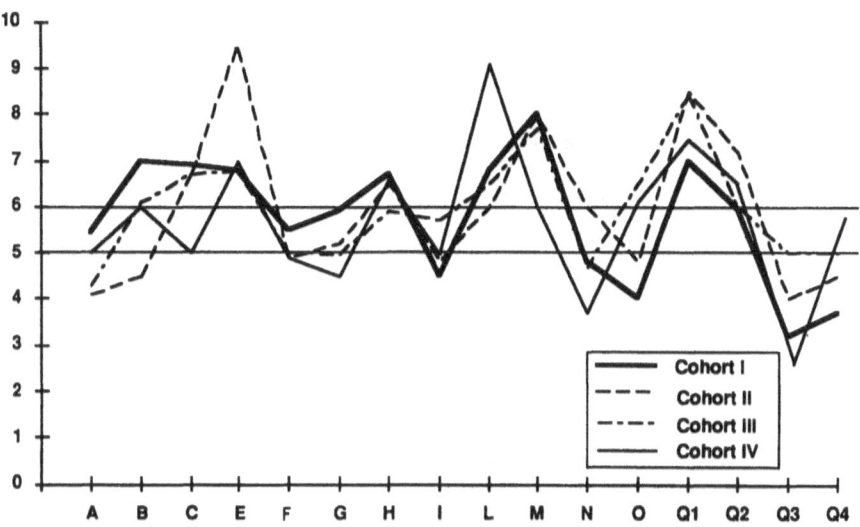

Fig. A.1. Personality profiles by cohort (16PF)

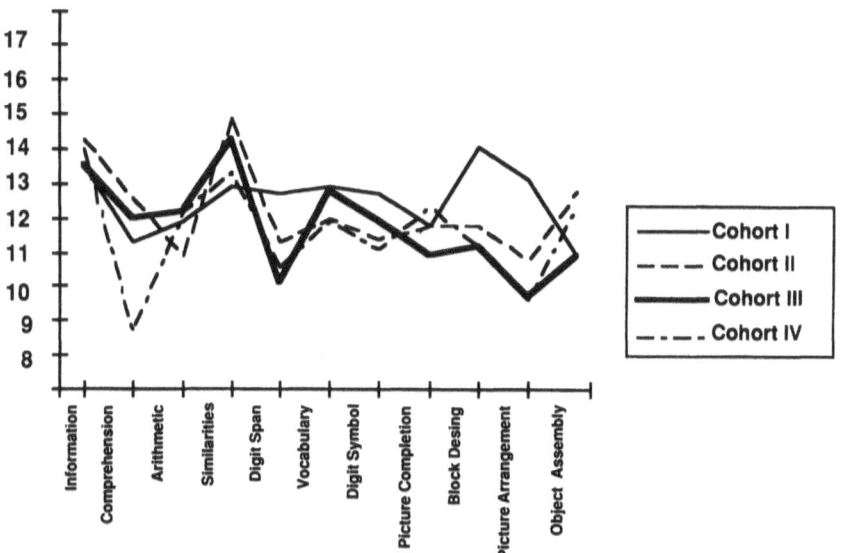

Fig. A.2. Intelligence subtest profiles by cohort (WAIS)

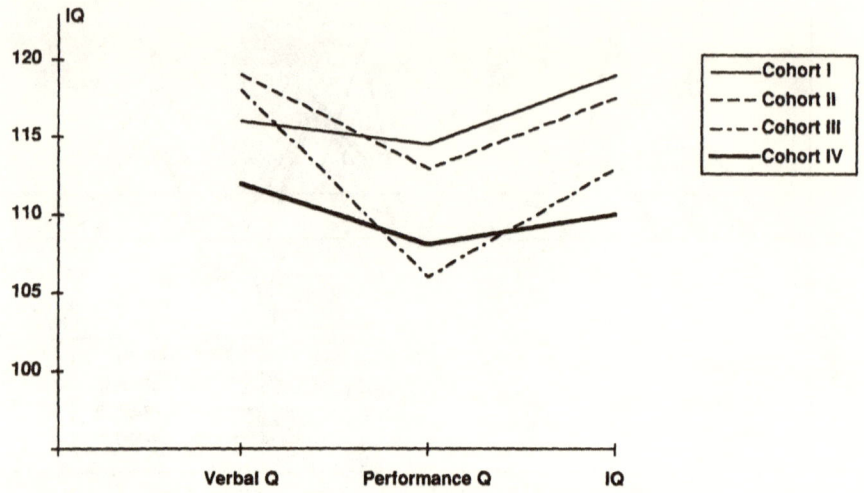

Fig. A.3. Intelligence profiles by cohort (WAIS)

Fig. A.4. Distribution of most-socialized or least-socialized students according to a canonic discriminating function

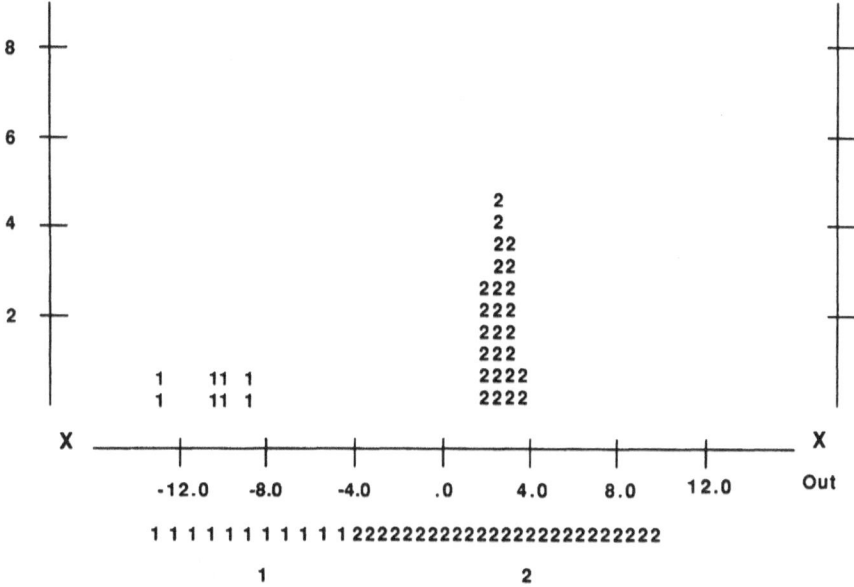

1 = Most socialized students
2 = Least socialized students

Bibliography

Aberle, D. F. 1961. Culture and Socialization. In *Psychological Anthropology: Approaches to Culture and Personality*, ed. F.L.K. Hsu, 381–97. Homewood, Ill.: Dorsey Press.
Acosta, Mariclaire, Jorge Bartolucci, and Roberto A. Rodríguez. 1981. *Perfil del alumno de primer ingreso al Colegio de Ciencias y Humanidades*. Mexico: CCH-UNAM.
Althusser, Louis. 1976. Ideología y aparatos ideológicos del estado. In *La filosofía como arma de la revolución*, 97–145. Cuadernos y Pasado de Presente 4. Mexico: Siglo XXI.
Amsterdamska, Olga. 1987. *Schools of Thought: The Development of Linguistics from Bopp to Saussure*. Dordrecht and Boston: D. Reidel Publishing.
Bachelard, Gaston. 1979. *La formación del espíritu científico*. 7th ed. Mexico: Siglo XXI.
Barber, Bernard. 1952. *Science and the Social Order*. New York: The Free Press.
Barnes, S. B., and R.G.A. Dolby. 1970. The Scientific Ethos: A Deviant Viewpoint. *Archives Européenes de Sociologie* 9(1):3–25.
Barron, Frank. 1965. The Psychology of Creativity. In *New Directions in Psychology II*. London: Holt, Rinehart & Winston.
Becker, Howard, B. Geer, E. Hughes, and A. Strauss. 1961. *Boys in White: Student Culture in Medical School*. Chicago: University of Chicago Press.
Ben-David, J. 1962. Universities and Academic Systems in Modern Societies. *European Journal of Sociology* 3:45–84.
———. 1968. *Fundamental Research and the Universities*. Paris: OECD.
———. 1971. *The Scientist's Role in Society*. Englewood Cliffs, N.J.: Prentice-Hall.
Berger, Peter, and Thomas Luckman. 1976. *The Social Construction of Reality*. London: Penguin Books.
Bernal, John D. 1979. *Science in History*. 4 vols. Cambridge: MIT Press.
Bernard, Russell H., and P. D. Killworth. 1977. Scientists as Others See Them. *Ocean Development and International Law Journal* 4(3):261–68.
Bleichmar, Hugo B. 1976. *La depresión. Un estudio psicoanalítico*. Buenos Aires: Ediciones Nueva Visión.
Bock, Phillip. 1969. *Modern Cultural Anthropology*. New York: Alfred A. Knopf.
Bourdieu, Pierre. 1988. *Homo Academicus*. Stanford, Calif.: Stanford University Press.
Bravo Ugarte. 1947. *Historia de México. La Nueva España*. Vols. 1 and 2. Mexico: IUS.
Bucher, Rue, and Joan G. Stelling. 1977. *Becoming Professional*. Library of Research 46. Beverly Hills: Sage.

Bunge, Mario. 1991. A Critical Examination of Science. *Philosophy of the Social Sciences* 21(24):524–60.
Camp, Roderic A. 1976. The University and Political Leadership in Mexico. Mimeo.
Cannon, Walter B. 1965. *The Way of an Investigator*. New York: Hafner Publishing Co.
Carvajal, Raúl, and Larissa Lomnitz. 1981. El desarrollo científico en México ¿Es posible multiplicarlo con los mismos recursos? *Ciencia y Desarrollo* 7(37):90–98.
Castañeda, Mario. 1974. Una carrera de investigación biomédica. In *Los perfiles de la bioquímica en México*, ed. Castañeda, M.J. Mora, S. Estrada O., and J. Martuscelli, 515–28. Mexico: UNAM.
Castañeda, Mario, J. Martuscelli, and J. Mora y Negrete. 1975. Las crisis de identidad en el científico. In *México: Ciencia y desarrollo*, 9–14.
Cattell, R. B. 1945. Principal Trait Clusters for Describing Personality. *Psychological Bulletin* 42:129–61.
Cattell, R. B., and H. W. Eber. 1962. *Sixteen Personality Factor Questionnaire. Manual for Forms A and B*. Champaign, Ill.: Institute for Personality and Ability Testing.
Cattell, R. B., H. W. Eber, and M. M. Tatsuoka. 1970. *Handbook for the Sixteen Personality Factor Questionnaire (16PF)*. Champaign, Ill.: Institute for Personality and Ability Testing.
Cereijido, Marcelino. 1991. *La nuca de Houssay*. Buenos Aires: Fondo de Cultura Económica.
Chubin, Daryl E. 1990. Scientific Malpractice and the Contemporary Politics of Knowledge. In *Theories of Science in Society*. Ed. Susan Cozzens and Thomas Gieryn. Bloomington: Indiana University Press.
Clark, Burton R., comp. 1980. *Academic Culture*. Working Paper (March). New Haven: Yale Higher Education Research Group.
Clausen, John A., comp. 1968. *Socialization and Society*. Boston: Little, Brown & Co.
Collins, H. M. 1985. *Changing Order: Replication and Induction of Scientific Practice*. London and Beverly Hills: Sage.
CONACYT. 1976. *Plan nacional indicativo de ciencia y tecnología*. Mexico: Consejo Nacional de Ciencia y Tecnología.
———. 1978. *Programa nacional de ciencia y tecnología 1978–1982*. Mexico: Consejo Nacional de Ciencia y Tecnología.
CONACYT-SEP. 1991. *Indicadores, actividades científicas y tecnológicas*. Mexico: CONACYT-SEP.
Coser, Lewis A. 1978. *Las instituciones voraces*. Mexico: Fondo de Cultura Económica.
Cozzens, Susan, and Thomas Gieryn, eds. 1990. *Theories of Science in Society*. Bloomington: Indiana University Press.
Crane, Diana. 1972. *Invisible Colleges: Diffusion of Science in Scientific Communities*. Chicago: University of Chicago Press.
Cravioto Magallón, R. M. 1971. Estudio de confiabilidad de la prueba de 16 factores de personalidad (16PF) del Dr. Raymond B. Cattell. Undergraduate thesis, UNAM.
Cruz Manjarrez, Héctor. 1982. Historia de la física en México y algunas de las ciencias afines. Mimeo.

Cueto, Marco S. 1990. The Rockefeller Foundation's Medical Policy and Scientific Research in Latin America: The Case of Physiology. *Social Studies on Science* 20(2):229–45.
De Gortari, Eli. 1979. *La ciencia en la historia de México*. Mexico: Editorial Grijalbo.
Dellas, M., and E. L. Gaier. 1970. Identification of Creativity: The Individual. *Psychological Bulletin* 73(1):55–73.
De María y Campos. 1980. *Estudio histórico-jurídico de la Universidad Nacional (1881–1929)*. Mexico: Dirección de Estudios y Programas Legislativos–UNAM.
———. 1981. Los científicos: Actitudes de un grupo de intelectuales porfirianos frente al positivismo y la religión. Paper presented at Sexta Reunión de Historiadores Mexicanos y Norteamericanos. Mimeo.
De Solla Price, Derek. 1979. *A Very Simple Model of the Revolution in Higher Education*. New Haven: Yale University Press.
De Vos, George. 1979. Apprenticeship and Paternalism. In *The Heritage of Endurance*, ed. G. de Vos and H. Wagatrina, 211–27. Berkeley and Los Angeles: University of California Press.
Doran, Chris. 1989. "Jumping Frames: Reflexivity and Recursion in the Sociology of Science." *Social Studies on Science* 19(3):515–35.
ECLAC-UNESCO. 1992. Educación y conocimiento: Eje de la transformación productiva con equidad. LC/G 1702 (SES 24/4) Rev. 1. Santiago de Chile: United Nations.
Elkin, F. 1960. *The Child and Society: The Process of Socialization*. New York: Random House.
Festinger, Leon. 1957. *A Theory of Cognitive Dissonance*. Stanford: Stanford University Press.
Freud, Sigmund. 1914. Introducción al narcisismo. In *Obras completas*, 3rd ed., vol. 2, 2017–2033. Madrid: Biblioteca Nueva.
———. 1920–21 [1921] Psicología de masas y análisis del Yo. In *Obras completas*, 3rd ed., vol. 3, 2563–2610. Madrid: Biblioteca Nueva.
Fuenzalida, Edmundo. 1971. Problemas de ciencia y tecnología en el paso al desarrollo. *Mensaje* 202–3:444–51.
———. 1972. La dependencia en el ámbito del saber superior y la transferencia de modelos universitarios en América Latina. *Jornadas Universitarias de CPU* 3:327.
Gaillard, Jacques. 1991. *Scientists in the Third World*. Lexington: University Press of Kentucky.
García Salord, Susana, and Liliana Vanella. 1992. *Normas y valores en el salón de clases*. Mexico: Siglo XXI.
García Stahl, Consuelo. 1975. *Síntesis histórica de la Universidad de México*. Mexico: Secretaría de Rectoría, Dirección General de Orientación Vocacional, UNAM.
Getzels, J. W., and P. W. Jackson. 1973. The Highly Intelligent and the Highly Creative Adolescent: A Summary of Some Research Findings. In *Scientific Creativity: Its Recognition and Development*, ed. C. W. Taylor and F. Barron, 161–72. John Wiley & Sons.
Giddens, Anthony. 1973. *The Class Structure of Advanced Societies*. New York: Harper & Row.
Goffman, W. 1966. Mathematical Approach to the Spread of Scientific Ideas: The History of Most Cell Research. *Nature* 212:449–52.

Gough, H. G. 1961. Techniques for Identifying the Creative Research Scientists. Paper presented at Conference on the Creative Person. University of California, Berkeley, Institute of Personality Assessment and Research.

Graciarena, Jorge. 1971. Modernización universitaria y clases medias: El caso de Brasil. In *Modernización y democratización en América Latina*, 93–142. Promoción Universitaria 3. Santiago de Chile.

Grene, Marjorie. 1984. Toward a New Philosophy of Science. In *Evolution at a Crossroads: The New Biology and the New Philosophy of Sciences*, ed. D. Depwe and B. Weber. Cambridge: MIT Press.

Grinberg L., and R. Grinberg. 1980. *Identidad y cambio*. Buenos Aires: Paidós.

Hackett, Edward. 1990. Science as a Vocation in the 1990s: The Changing Organizational Culture of Academic Science. *Journal of Academic Science* 61(3):241–79.

Hagendijk, Rob. 1990. Structuration Theory, Constructivism and Scientific Change. In *Theories of Science in Society*, ed. Susan Cozzens and Thomas Gieryn. Bloomington: Indiana University Press.

Hagstrom, W. O. 1975. *The Scientific Community*. Carbondale: Southern Illinois University Press, Arcturus Books.

Hull, L.W.H. 1973. *Historia y filosofía de la ciencia*. Barcelona: Editorial Ariel.

IIBM (Instituto de Investigaciones Biomédicas). 1977. *Licenciatura, maestría y doctorado en investigación biomédica básica*, 3–22. Mexico: Unidad Académica de los Ciclos Profesional y de Posgrado, UNAM.

———. 1981. *XL Aniversario del Instituto de Investigaciones Biomédicas*. Mexico: UNAM.

———. n.d.

Kaplan, Marcos. 1981. Bloqueos sociopolíticos a la ciencia y a la universidad en América Latina. *Vuelta* 54:30–35.

Karson, S., and J. W. O'Dell. 1987. *16PF guía para su uso clínico*. Madrid: Tea Editores.

Kerr, Clark. 1989. The Academic Ethic and University Teachers: A Disintegrating Profession? *Minervos* 27(2–3):139–56.

Knorr-Cetina, D. 1982. Scientific Communities or Transepistemic Arenas of Research? A Critique of Quasi-Economic Models of Science. *Social Studies of Science* 12(1):101–30.

———. 1983. The Ethnographic Study of Scientific Work: Toward a Constructionist Interpretation of Science. In D. Knorr-Cetina and Michael Mulkay, eds., *Science Observed*. London: Sage.

Koestler, A. 1975. *The Art of Creation*. 3rd ed. New York: Dell Publishing Co.

Kohut, H. 1980. *La restauración de sí-mismo*. Buenos Aires: Paidós.

Kuhn, T. S. 1962. *The Structure of Scientific Research*. Chicago: University of Chicago Press.

———. 1977. The Essential Tension: Traditional Innovation in Scientific Research. In *Scientific Creativity: Its Recognition and Development*, ed. C. W. Taylor and F. Barron, 225–39. Chicago: University of Chicago Press.

Lacan, J. 1976. La ciencia y la verdad. In *Escritos*. Vol. 1. Mexico: Siglo XXI.

Lacey, Colin. 1977. *The Socialization of Teachers*. London: Methuen.

Laing, R. D. 1969. *Self and Others*. New York: Pantheon Books.

Laplanche, J., and J. B. Pontalis. 1974. *Diccionario de psicoanálisis*. Barcelona: Labor.

Latour, B., and S. Woolgar. 1978. *Laboratory Life: The Social Construction of Scientific Facts*. London: Sage.

Leff, E. 1979a. Dependencia científico-tecnológica y desarrollo económico. In *México hoy*, comp. P. González Casanova and E. Florescano, 267–83. Mexico: Siglo XXI.
———. 1979b. Investigación científica e investigación tecnológica. In *La Universidad Nacional y los problemas nacionales*, vol. 2, 243–46. Mexico: Sociedad y Cultura–UNAM.
Lévi-Strauss, Claude. 1968. The Effectiveness of Symbols. In *Structural Anthropology*, 186–205. London: Penguin Books.
———. 1969. The Story of Asdiwal. In *The Structural Study of Myth and Totemism*, ed. Edmund Leach, 1–47. ASA Monographs 5. London: Tavistock Publications.
Lomnitz, Larissa. 1972. *Estructura de la organización social de un instituto de investigación. Documento de trabajo*. Mexico: Coordinación de Ciencias–UNAM.
———. 1976. La antropología de la investigación científica en la UNAM. In *La ciencia en México*, comp. L. Cañedo and L. Estrada, 13–25. Mexico: Fondo de Cultura Económica.
———. 1977. Conflict and Mediation in a Latin American University. *Journal of Interamerican Studies and World Affairs* 19(3):315–38.
———. 1979. Hierarchy and Peripherality: The Organization of a Mexican Research Institute. *Minerva* 4:527–48.
———. 1981. El congreso científico: Una perspectiva antropológica. *Vuelta* 59:45–48.
Lomnitz, Larissa, and Cinna Lomnitz. 1977. La creación científica. *Pensamiento Universitario* 3. Mexico: UNAM.
Lomnitz, Larissa, and Marisol Pérez-Lizaur. 1987. *An Elite Family of Mexico, 1812–1980*. Princeton: Princeton University Press.
López Cámara, Francisco. 1971. *El desarrollo de la clase media*. Mexico: Cuadernos de Joaquín Mortiz.
Lynch, Michael. 1985. *Art and Artifact in Laboratory Science*. London: Routledge & Kegan Paul.
Mahoney, M.J. 1979. Psychology of the Scientist: An Evaluative Review. *Social Studies of Science* 9(4):349–75.
Malinowski, Bronislaw. 1971. *Magic, Science and Religion*. New York: The Free Press.
Mead, George Herbert. 1934. *Mind, Self and Society*. Chicago: University of Chicago Press.
Merton, R. K. 1938. Science, Technology and Society in Seventeenth Century England. *Osirys* 4(2):[reprinted with a new introduction, New York: Harper & Row, 1970].
———. 1973a. Priorities in Scientific Discovery. In *The Sociology of Science*, 286–324. Chicago: University of Chicago Press [1st ed. 1942].
———. 1973b. The Puritan Spur to Science. In *The Sociology of Science*. ed. Jerry Gaston, 228–53. San Francisco: Bass Publishers.
———. 1976. The Ambivalence of Scientists: A Postscript. In *Sociological Ambivalence*, 32–55. New York: The Free Press.
Merton, R. K., G. G. Reader, and P. Kendall, eds. 1943. *The Student Physician: Introductory Studies in the Sociology of Medical Education*. Cambridge: Harvard University Press [1st ed. 1943].
Mitroff, Ian I. 1972. The Myth of Objectivity or Why Science Needs a New Psychology of Science. *Management Science* 18:613–18.

―――. 1974. *The Subjective Side of Science*. Amsterdam: Elsevier.
Moravezsik, M. J. 1980. *How to Grow Science*. New York: Universe Books.
Moreno de los Arcos, Roberto. 1975. La ciencia de la ilustración mexicana. *Anuario de Estudios Americanos* (Seville) 23:25–41.
Mulkay, M. J. 1969. Some Aspects of Cultural Growth in the Natural Sciences. *Social Research* 36:22–52.
―――. 1974. Methodology on the Sociology of Science: Reflections on the Study of Radio Astronomy. *Social Science Information* 13(2):107–19.
―――. 1976. The Mediating Role of the Scientific Elite. *Social Studies of Science* 6:295–342.
―――. 1977. Sociology of the Scientific Research Community. In *Science, Technology and Society*, ed. Ina Spiegel-Rosing and D. de Solla Price, 93–148. London: Sage.
―――. 1979. *Science and the Sociology of Knowledge*. London: Allen & Unwin.
―――. 1980. Interpretation and the Use of Rules: The Case of the Norms of Science. In *Transactions*, 111–25. New York: New York Academy of Sciences.
Nieto, D. A. 1981. Historia del Instituto de Investigaciones Biomédicas (1941–1965). In *XL Aniversario del Instituto de Investigaciones Biomédicas*, 9–12. Mexico: UNAM.
Otto, Richard. 1982. Sobre la naturaleza profesionalizante de la universidad. *Pensamiento Universitario* 56. Mexico: Centro de Estudios sobre la Universidad, UNAM.
Paradise, Ruth. 1978. The Teaching of Specific Attitudes and Values Which Aid in Adaptation of Future Work Situations: A Study of the Teacher–Student Interaction in Mexican Primary Schools. Paper presented at Annual Meeting of Society for Applied Anthropology, Mérida, Mexico, Sept. Mimeo.
Pavalko, Ronald M., and John M. Holley. 1974. Determinants of a Professional Self-Concept Among Graduate Students. *Social Quarterly* 55:462–77.
Paz, Octavio. 1986. *Sor Juana Inés de la Cruz o las trampas de la fe*. Mexico: Fondo de Cultura Económica.
Pérez Correa, Fernando. 1974. La universidad: Contradicciones y perspectivas. *Foro Internacional* 3:14–65.
Pérez Tamayo, Ruy. 1991. La ciencia en México. In *Ciencia, paciencia y conciencia en México*. Mexico: Siglo XXI.
Pinch, Trevor J. 1979. Normal Explanations of the Paranormal: The Demarcation Problem and Freud in Parapsychology. *Social Studies of Science* 9(3):329–48.
Prigogine, I., and I. Stengers. 1984. *Order Out of Chaos: Man's New Dialogue with Nature*. London: Bantam Books.
Reid Angu. 1981. Socialization into the Professions: The Impact of Dental School Faculty on Students' Professional Orientations. *The Canadian Review of Sociology and Anthropology* 19(1):48–66.
Reidl, L. M. 1969. Estudio preliminar a la estandarización de la prueba "JR-SR high school personality questionnaire" del Dr. Raymond B. Cattell en un grupo de adolescentes. Undergraduate thesis, UNAM.
Reinhartz, Shulamit. 1979. *On Becoming a Social Scientist*. San Francisco: Jossey Bass Publishers.
Ribeiro, Darcy. 1971. *La universidad latinoamericana*. Santiago de Chile: Ediciones Universitarias.

Rodríguez Sala de Gómez Gil, M. L., and A. Chavero González. 1982. *El científico en México: Su formación en el sistema extranjero, su incorporación y adecuación al sistema ocupacional mexicano.* Mexico: Instituto de Investigaciones Sociales, Centro de Estudios sobre la Universidad, UNAM.
Rose, Hilary, and Steven Rose. 1969. *Science and Society.* London: Penguin Books.
Rosenblueth, Arturo. 1971. *El método científico.* Mexico: Centro de Investigaciones y Estudios Avanzados del Instituto Politécnico Nacional.
Schoijet, Mauricio. 1979. The Condition of Science in Mexico. Mimeo.
Scott, Robert. 1968. Student Political Activism in Latin America. *Daedalus.*
Sears, Robert R., E. E. Maccoby, and H. Levin. 1957. The Socialization of Aggression. In *Readings in Social Psychology.* New York: Rinehart & Winston.
Silva Herzog, Jesús. 1974. *Una historia de la Universidad de México y sus problemas.* Mexico: Siglo XXI.
Silva Michelena, Héctor, and H. R. Sonntag. 1980. *Universidad, dependencia y revolución.* Mexico. Siglo XXI.
Smith, Peter. 1975. La movilidad política en el México contemporáneo. *Foro Internacional* 15:379–413.
Snow, N. 1965. *The Search.* London: Penguin Books.
Stepan, Nancy. 1976. *Beginnings of Brazilian Science.* New York: Science History Publications.
Stolte-Haiskanen, Veronica. 1989. Cultural Repertoires of the Scientific Community and Problems of Legitimation of Science in Society. In *J. D. Bernal's "The Social Function of Science,"* ed. H. Steinder, 415–34. Berlin: Akademie-Verlag.
Storer, Norman. 1966. *The Social System of Science.* New York: Holt, Rinehart & Winston.
Szent-Gyorgy, A. 1962. Scientific Creativity. *Perspectives in Biology and Medicine* 4:173–78.
Tapia, Ricardo. 1989. Los institutos y centros como sedes de licenciaturas y posgrados de investigación biomédica básica de la UACP y P del CCH. Paper presented at UNAM's "Foros de Discusión."
Tedesco, Juan Carlos. 1971. Modernización y democratización en la Universidad Argentina: Un panorama histórico. In *Modernización y democratización en América Latina,* 143–87. Promoción Universitaria 3. Santiago de Chile.
Trabulse, Elías. 1983. *Historia de la ciencia en México.* Mexico: CONACYT/Fondo de Cultura Económica.
Trow, Martin. 1970. *The Expansion and Transformation of Higher Education.* Berkeley: University of California, for General Learning Corporation. Mimeo.
UNAM. 1990. *Dirección General de Planeación.* Agenda estadística.
Van de Graaf, J. H., B. R. Clark, D. Furth, D. Goldsmidt, and D. F. Wheeler. 1978. *Academic Power.* New York: Praeger.
Vessuri, Hebe. "Perspectivas en el Estudio Social de la Ciencia." *Interciencia* 16(2):60–68.
Von Foerster, Heinz. 1981. *Observing Systems.* San Francisco: Intersystems Publications.
Wallerstein, Immanuel. 1974. *The Modern World System I.* New York: Academic Press.
Watson, J. D. 1968. *The Double Helix.* New York: Atheneum.
Watzlawick, Paul, Janet Helmick Beavin, and Don D. Jackson. 1971. *Teoría de la comunicación humana.* Buenos Aires: Editorial Tiempo Contemporáneo.

Weber, Max. 1949. *La ética protestante y el espíritu del capitalismo*. Buenos Aires: Editorial Díaz.
———. 1974. *The Methodology of Social Sciences*. New York: The Free Press.
Willms, Kaethe. 1981. Memorias. In *XL Aniversario del Instituto de Investigaciones Biomédicas*, 13–17. Mexico: UNAM.
Winnicot, D. W. 1972. *Realidad y juego*. Buenos Aires: Granica Editor.
Ziman, John D. 1980a. Rights and Responsibilities in Research: Dichotomy of Dynamic Balance. *Society and Science* 3(4):109–16.
———. 1980b. What Are the Options: Social Determinants of Personal Research Plans. Paper presented at the Conference on Scientific Establishments and Hierarchies, Oxford, July 5 [publication forthcoming in *Minerva*].
———. 1981. Social Responsibility of Scientists: Basic Principles. Mimeo.
Zinberg, Dorothy S. 1974. Science in a Social Psychological Activity. In *Social Process of Scientific Development*, 242–53. London: Routledge & Kegan Paul.
Zuckerman, Harriet. 1977. *Scientific Elite: Nobel Laureates in the United States*. New York: The Free Press.
———. n.d. Stratification in American Science. *Sociological Inquiry* 40:235–57.

Index

Academy for Scientific Research, 18
admission to UNAM
   requirements for, 40–42
   and socialization, 101
advisory system, 34, 67. *See also* teaching methods, at UNAM
Althusser, Louis, 75
Alzate y Ramírez, José, 15
ambivalence (concept), 88, 92
Apollo mission study, 91
apprenticeship, 153
Astronomical Observatory, Chapultepec, 16, 18
Ateneo de la Juventud (Atheneum of Youth), 17
authority vs. reason, 12

Barber, Bernard, 91
Barnes, S. B., 92
Barreda, Gabino, 16, 24
Bartolache, J. I., 15
Ben-David, J., 19, 21
Berger, Peter, 153
Bernal, John D., 12
Bernard, Russell H., 76–77
bibliographic research, 63, 65, 66–67, 73
bioengineering curriculum, 176, 177
biology curriculum. *See* curriculum, at UNAM, in biology
Bourdieu, Pierre, 7n.1
brochure for UNAM program, 100–101
Bucher, Rue, 159, 160
budgets, at UNAM, 26. *See also* financial support for science

Calvinism, 12
capitalism, 12

Cattell, Raymond B., 182. *See also* 16PF personality test
CCH (Colleges of Sciences and Humanities), 26, 38, 40n.4, 42–43
Centro de Investigaciones Ecológicas del Sureste (CIES, or Chiapas regional center), 44, 178
Chapultepec Astronomical Observatory, 16, 18
Chávez, Ezequiel, 16, 24
chemistry curriculum, 169, 171, 173, 180
CIES (Centro de Investigaciones Ecológicas del Sureste, or Chiapas regional center), 44, 178
Clavijero, Francisco Javier, 14
cognitive flexibility, 85
College of Mining (formerly Seminary of Mining), 15
Colleges of Sciences and Humanities (CCH), 26, 38, 40n.4, 42–43
Collins, H. M., 55n.1
commitment
   in ideology of science, 77, 83, 91
   socialization for, 105
communism, 90
competition, 83, 87
Comte, Auguste, 16
CONACYT (National Council of Science and Technology), 3, 18
congresses, in socialization, 122–24, 137
constructivism, 93–94
control structure, 78–83, 88–89
convergent thought, definition of, 88
cooperation, in motivation, 87
Córdoba movement, 22
Coser, Lewis A., 163
cosmology, 76–77

220 Index

creativity
  definition of, 85
  and emotions, 81
  in ideal scientist model, 77, 78–79, 85–86
  students' modeling of, 124
  teaching methods for developing, 57, 67, 72
critical reading, 62–63
curiosity (inquisitiveness), 42
curriculum, at UNAM's experimental basic biomedical research program
  overview of, 40, 165–68
  1st year, 169–73
  2nd year, 173–76
  3rd year, 176–79
  4th year, 179–81
  in bioengineering, 176, 177
  in biology: 1st year, 170, 171, 173; 2nd year, 173, 175; 3rd year, 177, 179; 4th year, 180
  in chemistry, 169, 171, 173, 180
  in immunology, 177–78, 181
  in mathematics, 169, 170
  in medicine in general, 179–80, 181
  in parasitology, 177, 178
  in pathology, 167, 176, 177, 178, 181
  in physiology, 173, 175–76
  in statistics, 171, 173
  in technical skills: overview of, 166–67, 168, 170, 171; 1st year, 170, 171–72; 2nd year, 173–74, 175, 176; 3rd year, 177, 178; 4th year, 180
  in theoretical knowledge, 179, 180

debate, rules for, 120. See also discussion
degree coordinators, 39
delayed gratification, 82, 87
del Río, Andrés Manuel, 15
developing countries, conditions in, 2, 160–61, 162
Díaz, Porfirio, 16–17
Directorate of Geographic and Climatological Studies, 18
discipline
  emotional controls, 78, 81–83
  in ideal scientist model, 77, 78–83
  mental discipline, 78, 79–81
  work discipline, 78, 79, 84–85
discussion
  in socialization, 118–21, 126–28
  teaching methods for, 57–58, 62–63, 72–74, 119, 126

disinterest, 90, 91, 92
disorder vs. order, 89
doctoral work
  and identity formation, 158
  statistics on, 18–19, 164
Dolby, R. G. A., 92

Economic Commission for Latin America and the Caribbean (ECLAC), 2–3
ecopathology, 167
elementary education, 17
emotions
  control of, 78, 81–83
  release of, 78, 86–89
ENEP (Schools for Professional Studies), 26
Enlightenment, in Mexico, 13–15, 17
entrepreneurs, children of, 43
Escuela Nacional Preparatoria (National Preparatory School), 16, 24
essential tension (concept), 88–89
ethical norms, 90. See also ideology, of science
ethos, 90
evaluation of students, 113–15, 160
expenditures for science, 3, 4, 5, 6, 8, 9
experimental (problem solving) work. See National Autonomous University of Mexico (UNAM) undergraduate program, experimental (problem solving) work in

financial support for science, 3, 4, 5, 6, 8, 9
Freud, Sigmund, 148
frustration, tolerance for, 82–83

General Hospital, 44, 48
generalizability, 91
González Guzmán, Ignacio, 31
Gough, H. G., 85
graduate degrees, statistics on, 18–19, 20
gratification, delayed, 82, 87
group work/interaction
  in motivation, 87
  networks in, 99, 118, 152
  in socialization: discussion groups, 118–21, 126–28; in fourth year program, 139, 152; in general, 97, 99, 117–18; and identity, 151–52; symposiums/congresses, 122–24; in third year program, 136–37, 152; work groups, 121–22

Guevara, Andrés de, 15
Gutiérrez, Alonso, 13

Hackett, Edward, 94
Hagstrom, W. O., 159–60
high schools, 16
honesty (integrity), 83, 91
Hospital del Niño, 179
Humboldt Scientific Society, 16

ideal scientist, model of
  control structure (discipline in), 77, 78–83, 88–89
  function of, 75–76
  and identity formation, 150, 154, 156, 157, 158
  liberating processes in, 78–79, 80, 83–89
  and overidealization, 162–63
  overview of, 77–79
  and socialization in general, 133, 149, 162–63
  See also scientific identity, formation of
identity, 146–48. See also scientific identity, formation of
ideology, nationalist, 48, 133–34
ideology, of science
  cosmology of, 76–77
  in developing countries, 161
  general importance of, 75, 76
  review of previous research on, 90–94
  and socialization, 94–95, 97, 116, 123, 140, 151, 159, 162
  values in, 77–83, 90, 91–92, 95, 162
  See also ideal scientist, model of; scientific identity, formation of
imagination. See creativity
immunology curriculum, 177–78, 181
independence (personality characteristic), 78, 88, 106–10, 149–50
independence, of Mexico, 15, 17
individualism, 12, 84, 91
industrialization, in Mexico, 16, 22
information acquisition
  mental discipline for, 79–80
  socialization for, 124–25
  teaching methods for, 56–57, 62
inquisitiveness (curiosity), 42
Institute of Biomedical Research (formerly Laboratory of Medical and Biological Studies), 31–32, 33, 34–36, 38, 39–40, 44, 46–47

institutes, research. See research institutes
Instituto Politécnico Nacional (National Polytechnic Institute), 18
integrity (honesty), 83, 91
intelligence
  as criterion for admission, 41
  in ideology of science, 91
  statistical analysis of students' narrative discussion of, 185–89, 190; tabular results of, 196–98, 200, 202, 204, 205, 207, 208
Internal Council, 39, 46
introductory courses, 41, 42, 102

Jesuits, 14
Jewish students, 43n.6
Juárez, Benito, 24

Killworth, P. D., 76–77
Kuhn, Thomas S., 88–89, 92

Laboratory of Medical and Biological Studies (now Institute of Biomedical Research), 31–32, 33, 34–36, 38, 39–40, 44, 46–47
laboratory work in socialization, 129–33. See also National Autonomous University of Mexico (UNAM) undergraduate program, experimental (problem solving) work in
language of science, 79–80
Latour, B., 94
León y Gama, Antonio, 15
liberating processes, 78–79, 80, 83–89
logical reasoning, 80, 81
Luckman, Thomas, 153

Mahoney, M. J., 91
manual skills, learning of. See techniques, learning of
mathematics curriculum, 169, 170
Maximilian, emperor of Mexico, 16
medicine, curriculum in, 179–80, 181
mental discipline, 78, 79–81. See also emotions, control of
Merton, R. K., 88, 90–94, 159–60, 162
methodological norms, 90–91
methodology, for study of UNAM program, 3–7
methodology, learning of. See techniques, learning of

Military Hospital, 44, 48
Mitroff, Ian I., 91, 94
Mora, Jaime, 32
moral norms, 90. *See also* ideology, of science
motivation, as criteria for admission, 41–42
motivational stimuli, 78, 82, 87–88
Mulkay, M. J., 91–93
myth and science, 77. *See also* ideal scientist, model of

National Autonomous University of Mexico (UNAM)
  budget of, 26
  establishment of, 24–25
  history of scientific study at, 28, 31–32
  Institute of Biomedical Research, 31–32, 33, 34–36, 38, 39–40, 44, 46–47
  number of researchers at, history of, 18–19
  number of students at, 26, 27, 141, 142, 164
  politics vs. academics at, 26, 28
  See also National Autonomous University of Mexico (UNAM) undergraduate basic biomedical research program
National Autonomous University of Mexico (UNAM) undergraduate basic biomedical research program
  1st year program: and identity formation, 153–55; introductory course, 41, 42, 102; overview of, 49–52; teachers of, 47–48, 49; teaching methods for, 55–67; time schedule for, 44–45
  2nd year program: and identity formation, 155; teachers of, 48; teaching methods for, 67–72; time schedule for, 45
  3rd year program: off-campus courses in, 48, 133–35; socialization during, 133–37; teachers of, 48–49, 134–35
  4th year program: and identity formation, 157–58; socialization during, 53, 137–40; teachers of, 49
  admission requirements for, 40–42
  experimental (problem solving) work in: scheduling of, 50; and socialization, 99, 107–9, 129–33; teaching methods for, 59–62, 63–67, 73
  facilities for, 44
  goals of, 36–37, 161–62
  introductory course, 41, 42, 102

methodology for study of, 3–7
number of students in/graduates of, 141, 142, 164
organizational structure of, 39–40
origins of, 35–38
time schedule for, 44–46, 129
See also curriculum; socialization at UNAM; students; teachers; teaching methods
National Council of Science and Technology (CONACYT), 3, 18
nationalism, in teaching methods, 48, 133–34
National Polytechnic Institute (Instituto Politécnico Nacional), 18
National Preparatory School (Escuela Nacional Preparatoria), 16, 24
National System of Researchers (SNI), 3, 7
National System of Science and Technology (SINCYT), 5, 6, 9
networks, in socialization, 99, 118, 152
Neurology Hospital, 181
norms, in scientific ideology, 90–95
Nutrition Hospital, 32

objectivity, 82, 90, 91
off-campus courses, 48, 133–35
order vs. disorder, 89
organized skepticism, 90, 92

parasitology curriculum, 177, 178
pathology curriculum, 167, 176, 177, 178, 181
personality, statistical analysis of students'
  narrative discussion of, 183–85, 190–92
  tabular results of, 193–95, 199, 201, 203, 205, 207
Physica speculatio (Gutiérrez), 13
physiology curriculum, 173, 175–76
politics and UNAM, 26, 28
Pontifical University of Mexico (Real y Pontificia Universidad de la Nueva España), 13, 15, 24
positivism, 16, 17, 24
Prigogine, I., 77–78n.1
problem solving work. *See* National Autonomous University of Mexico (UNAM) undergraduate basic biomedical research program, experimental (problem solving) work in
Puritan ethic, 90

Ramón y Cajal, Santiago, 31
Ramón y Cajal Institute, 31
rationality, 91, 92
reading, critical, 62–63
Real y Pontificia Universidad de la Nueva España, 13, 15, 24
reason vs. authority, 12
reconstruction, postrevolutionary, 17–19
recruitment, in socialization, 100–101
Reformation, 11, 17
Reid, Angu, 160
Reinhartz, Shulamit, 159
religion, in history of science, 13, 14–15
Renaissance, European, 11–12
research, bibliographic, 63, 65, 66–67, 73
research careers, randomness in choice of, 34, 36–37
research design, learning of, 67, 71–72, 180
researchers, student image of, 53
research institutes
  history of, 18, 22, 23, 28
  Institute of Biomedical Research, 31–32, 33, 34–36, 38, 39–40, 44, 46–47
  Ramón y Cajal Institute, 31
  UNAM budgets for, 26
research topics, choice of, 35, 37
research vs. teaching, 1, 21
Revolution, 1910, 16, 17, 18, 25
rewards for students, 113–15
Rockefeller Foundation, 31
role playing in socialization
  and discussion seminars, 126–28
  in general, 97–98, 99
  and identity formation, 148, 149, 155–56, 160
  and information acquisition vs. creativity, 124–25
  and laboratory work, 129–33
  and reading assignments, 125–26
  teachers as role models, 117, 145–46, 149

schedules, for UNAM undergraduate program, 44–46, 129
Scholasticism, 11, 12, 13, 15
School of Chemistry, 18
school of thought, definition of, 32n.2
Schools for Professional Studies (ENEP), 26
science, historical development of
  in Mexican universities, 21–24, 28, 31–32
  in Mexico in general, 2–3, 13–19, 28
  worldwide, 11–12

scientific congresses, in socialization, 122–24, 137
scientific identity, formation of
  definition of scientific identity, 146
  general discussion of, 145–46, 158–60
  and ideal scientist model, 150, 154, 156, 157, 158
  other research on, overview of, 158–60
  phases of: overview of, 151–53; 1st (powerlessness), 153–55; 2nd (overevaluation), 155–57; 3rd (questioning), 157; 4th (integration), 157–58; 5th (consolidation), 158
  and teaching methods, 149–51, 160
scientific ideology. See ideology, of science
scientific method, 77, 80–81. See also research design, learning of
scientist, ideal. See ideal scientist, model of
secondary schools, 16
self-confidence, 88
self-protecting feelings, 78, 88–89
seminars. See discussion
Seminary of Mining (now College of Mining), 15
Sierra, Justo, 16, 24–25
Sigüenza y Góngora, Carlos de, 14
SINCYT (National System of Science and Technology), 5, 6, 9
16PF personality test
  narrative discussion of, 182, 183–85, 190–92
  tabular results of, 193–95, 199, 201, 203, 205, 207
skepticism, organized, 90, 92
SNI (National System of Researchers), 3, 7
Soberón, Guillermo, 32, 38
social class, of students, 42–43
socialization. See also socialization at UNAM's experimental basic biomedical research program
  general discussion of, 143–45, 153
  and ideology of science, 94–95, 97, 116, 123, 151, 159, 162
  and liberating processes, 84
socialization at UNAM
  in 1st year program: role playing in, 126, 129, 130–32; teachers in, 51–52, 103–5, 109, 116
  in 2nd year program: during 2nd year program, 52–53; group interaction in, 123–24; and identity formation, 155; role playing in, 127, 129, 132; teachers

224 Index

in, 52–53, 104, 105, 109, 110–12, 116–17
in 3rd year program, 133–37
in 4th year program, 53, 137–40
and ideology of science, 116, 123, 140
initial contacts in, 100–102
and off-campus courses, 133–35
program goals for, 36
See also group work/interaction, in socialization; role playing in socialization; scientific identity; teaching methods, at UNAM, for socialization
Sor Juana Inés de la Cruz, 14
statistics curriculum, 171, 173
Stelling, Joan G., 159, 160
Storer, Norman, 91
students, at UNAM's experimental basic biomedical research program
class participation by, 57, 62–63
classroom relationships with teachers, 57–58, 102–5, 106–7, 110–12, 115–16, 149–51
demands on, 105–6, 153–55
evaluation of, 113–15, 160
in first year, overview of experiences of, 49–52
image of researchers held by, 53
numbers of, 26, 27, 141, 142, 164
social characteristics of, 42–43
statistical analysis of traits of: intelligence, 185–89, 190; methodology for, 182–83; personality, 183–85, 190–92; and socialization level, 189–92, 206, 209
See also scientific identity, formation of; socialization at UNAM
superego, 75, 90, 148
supervision of students, 110–12, 156
symposiums, in socialization, 122–24

teachers, at UNAM's experimental basic biomedical research program
of 1st year program, 47–48, 49
of 2nd year program, 48
of 3rd year program, 48–49, 134–35
of 4th year program, 49
academic traits of, 46–49
classroom relationships with students, 57–58, 102–5, 106–7, 110–12, 115–16, 149–51
of off-campus courses, 48, 134–35
in undergraduate admissions, 42
See also teaching methods, at UNAM

Teaching Committee, 41
Teaching Committee, 41
teaching methods, at UNAM's experimental basic biomedical research program
for 1st year program in general, 51, 55–67
for 2nd year program in general, 67–72
for 3rd year program in general, 48
for creativity development, 57, 67, 72
for critical reading, 62–63
for discussion, 57–58, 62–63, 72–74, 119, 126
early decisions on, 46–47, 55
goals for, 36
for identity formation, 149–51, 160
for information provision, 56–57, 62
nationalist ideology in, 48, 133–34
problem solving-based, 59–62, 63–67, 73
for research design learning, 67, 71–72
for socialization: in 1st year program, 51–52, 103–5, 109, 116; in 2nd year program, 52–53, 104, 105, 109, 110–12, 116–17; in 3rd year program, 134–36; in 4th year program, 138–39; demands on students in, 105–6, 153–55; for evaluation/rewards, 113–15; in general, 97–100, 102–5; importance of, other research on, 159–60; for independence development, 106–10; for supervision/guidance, 110–12, 156
and student participation, 57, 62–63
for technique (manual skills) learning, 58, 67–70, 72–73
for theory-based work, 63, 65–66, 73
See also curriculum, at UNAM
teaching vs. research, 1, 21
technical norms, 90–91
techniques, for study of UNAM program, 3–7
techniques, learning of
and socialization, 98
teaching methods for, 58, 67–70, 72–73
See also curriculum, at UNAM, in technical skills
tension, essential, 88–89
theory-based learning at UNAM
curriculum for, 179, 180
scheduling of, 50
teaching methods for, 63, 65–66, 73
thesis work, 74, 137–38, 180–81
Third World countries, conditions in, 2, 160–61, 162

time schedules, for UNAM's undergraduate basic biomedical research program, 44–46, 129
trust, 83

UNAM. *See* National Autonomous University of Mexico (UNAM)
Undergraduate Selection Committee, 41
universalism, 90
Universidad Obrera (Workers' University, formerly Universidad Gabino Barreda), 18
universities, in Mexico
  early history of, 13
  general characteristics of, 19–24
  in postrevolutionary period, 17–18
  Real y Pontificia Universidad de la Nueva España, 13
  Universidad Obrera, 18
  See also National Autonomous University of Mexico (UNAM)

values, in scientific ideology, 77–83, 90, 91–92, 95, 162. *See also* ideology, of science
Van de Graaf, J. H., 19, 21
Veláquez Cárdenas, Joaquín, 15
vocation, 151

WAIS intelligence test
  narrative discussion of, 182, 185–89, 190
  tabular results of, 196–98, 200, 202, 204, 205, 207, 208
Weber, Max, 6–7, 11
Wechsler, David, 182. *See also* WAIS intelligence test
work discipline, 78, 79, 84–85. *See also* discipline
Workers' University (Universidad Obrera, formerly Universidad Gabino Barreda), 18
work groups in socialization, 121–22

Zinberg, Dorothy S., 155

www.ingramcontent.com/pod-product-compliance
Lightning Source LLC
Chambersburg PA
CBHW031549300426
44111CB00006BA/237